Glendale College
Library

CHILD CARE:
A COMPREHENSIVE GUIDE

Model Programs and Their Components

A SERIES

edited by

Stevanne Auerbach, Ph.D.
with
James A. Rivaldo

Volume II

Companion Volumes in this series include:

Volume I. RATIONALE FOR CHILD CARE
SERVICES: PROGRAMS VS. POLITICS
Philosophy and historical perspective forming the basis of planning of model programs and services.

Volume III. CREATIVE CENTERS AND HOMES
Approaches to the design and implementation of the child's environment, and a focus on infant care and family day care homes.

Volume IV. SPECIAL NEEDS AND SERVICES
Methods for working with young children who are handicapped, abused, or having other special problems, and with an emphasis on parents, minority groups, and responsive programs.

CHILD CARE:
A COMPREHENSIVE GUIDE

VOLUME II
MODEL PROGRAMS AND THEIR COMPONENTS

Stevanne Auerbach, Ph.D., Editor
with
James A. Rivaldo

Foreword by

John Brademas
U. S. Representative, Indiana

HUMAN SCIENCES PRESS
SUBSIDIARY OF BEHAVIORAL PUBLICATIONS INC.
72 FIFTH AVENUE, NEW YORK, N.Y. 10011

Library of Congress Catalog Number 76–10121

ISBN: 0–87705–256–5

Copyright © 1976 by Stevanne Auerbach

HUMAN SCIENCES PRESS
72 Fifth Avenue
New York, New York 10011

Printed in the United States of America
6789 987654321

For all of the staff working in child care now and to
the students in training who will join them

CONTENTS

CONTRIBUTORS TO VOLUME II

KEITH R. ALWARD is a developer of competency units for early childhood education and a specialist in the theory and practical applications of Piaget. He is currently on the staff of the Far West Laboratory for Educational Research and Development.

SUSAN S. ARONSON, M.D. has been the project director for the Health Advocacy Training Project since 1973. This program serves 74 center and family day-care homes by providing on-site evaluation and consultation. Dr. Aronson is also executive director of the Learning Center, a day-care program operated by the Medical College of Pennsylvania. She also holds an assistant professorship in pediatrics and community and preventive medicine at the Medical College of Pennsylvania.

JEAN H. BERMAN is senior program analyst for child development at the Appalachian Regional Commission. Her previous experience in day-care licensing, early childhood education, and program development led to her particular involvement in the establishment of quality child-care centers in communities in rural areas.

DAVID BROWN, an elementary school teacher, has worked for the past ten years as a registered social worker in the Berkeley, California, Unified School District. Prior to that, he was employed by the American Friends Service Committee in community relations in an Indian-White community in California.

MARILYN CHOW is an assistant clinical professor in the Department of Family Health Care Nursing, University of California, San Francisco. She is teaching in the Maternal Child Nurse Associate Project. As part of a community pediatric fellowship of the National Urban Coalition, she received training and experience as a pediatric nurse associate; with that background, she was able to experiment and implement the role of the pediatric nurse associate in the day-care setting described.

HELEN L. GORDON, formerly a primary grade teacher in the Chicago public schools and elsewhere and a teacher of handicapped children in the public schools of several states, has also served extensively as a consultant and director to Head Start programs, parent-child programs, and day-care centers. Most recently she was executive director for the Portland Metro Community Coordinated Child Care Program. She has been a board member of the Day Care and Child Development Council of America.

CHRISTOPH M. HEINICKE (chairman), DAVID FRIEDMAN, ELIZABETH PRESCOTT, CONCHITA PUNCEL, and JUNE SOLNIT SALE make up the American Orthopsychiatric Association Study Group on the Mental Health Aspects of Day Care, which was activated by Dr. Jane Kessler, chairman of the Child Health Council of the association. This group first presented its findings in a panel session at the annual meeting of the association. They then published the paper, "The Organization of Day Care: Considerations relating to the Mental Health of Children and Family." In *American Journal of Orthopsychiatry*, January, 1973.

KATHLEEN BARRETT LATHAM was a project coordinator of Project Growing Together, at Gaywood Elementary Center, which served 70 families—primarily from the University of Maryland area. Previously, Ms. Latham taught preschool in Project Head Start in the District of Columbia and was the teacher-director of a co-op nursery school for two years. She is presently working on an M.Ed. in the Institute for Child Study at the University of Maryland in College Park, Maryland.

JIM LEVINE has been a preschool teacher, day-care director, and day-care consultant. He was a Ford Foundation travel-study fellow and wrote a book entitled *Men Who Care for Children.*

KAY MARTIN is a weaver, backpacker, and teacher. She has worked in community organizing for child care in Berkeley and is currently teaching a course at Berkeley, the University of California, entitled "Child Care: Theory, Policy and Practice."

MARY MILLMAN is a free-lance editor and a mother. She is currently a law student and a founding member of Bananas, a free child-care referral service in Berkeley. She has been an officer in the Berkeley Child Care Development Council.

ANN DeHUFF PETERS, M.D., is a pediatrician and public health physician with extensive training and experience in child-related areas. She has been a member of the faculties of Washington University, the University of North Carolina, and San Diego State University, and is at present clinical associate professor of pediatrics at the University of California, San Diego.

A fellow of the American Orthopsychiatric Association and the American Public Health Association and member of the National Association for the Education of Young Children and the Society for Research in Child Development, Peters has served as chairman of several task forces and committees dealing with day care. She has been consultant to the Committee on Infant and Preschool Child of the Academy of Pediatrics and to the State Departments of Health and Welfare in North Carolina and California. Dr. Peters was co-founder and medical director of a research demonstration day-care program for infants and very young children while at the University of North Carolina. She is at present working as an independent consultant.

LINDA REGELE-SINCLAIR has worked for The Children's Foundation, a Washington, D.C., based nonprofit organization which gives technical and organizing assistance to community groups, parents groups, etc., on the various federally sponsored child nutrition programs. At the foundation she has edited "Feed Kids," the monthly newsletter, and has written and prepared a widely distributed book entitled *Food Rights.* She also assists day-care and Head Start centers in obtaining assistance from USDA under the Special Food Service Program and has directed the foundation's work with the school lunch and breakfast program.

DOROTHY NASH SHACK, psychologist with the Early Growth Center of the Berkeley Unified School District, has also served as psychologist and reading clinician with reading and mental health clinics in the Chicago area. She has also been an instructor of psychology at Tennessee State University, Dillard University, and Haile Selassie University in Addis Ababa, Ethiopia.

MARY W. VAN VLACK, RAMON C. BLATT, AND PAUL T. BARNES have all actively participated in the evolution of the Counselling-Coordination Office (CCO). Ms. Van Vlack has been a research associate for the University of Colorado Medical Center Demonstration Child-

Care Project, which produced the CCO model; she is a sociologist by training and inclination. Dr. Blatt is currently director of the Project; he is a comparative psychologist by training, and sometimes by inclination. Dr. Barnes was director for the Child-Care Program; he is a developmental psychologist somewhat by training and always by inclination.

FOREWORD

Congressman John Brademas

Not long ago Kenneth Keniston, director of the Carnegie Council on Children, asked the question, "Do Americans *really* like children?"

His answer, reflecting his three years of work with the Council, was: "*Yes,* if our sentiments are to be taken as evidence."

"However," Keniston asserted, "in spite of our tender sentiments we do *not* really like children."

For, he asked, "Why *is* it that we, as a nation, allow so much inexcusable wretchedness among our children in practice, while at one and the same time we as individuals nurture and profess such tender and loving solicitous sentiments for our children?"

Keniston's explanation: ". . . our sentiments for children have been rendered ineffective by the stronger influences and forces of the economic system that have grown up willy-nilly among us."

Last year Senator Walter F. Mondale of Minnesota and

I introduced legislation, the Child and Family Services Bill, aimed at helping provide America's children the services that many of them require to live strong and healthy lives.

Passage of the bill—co-sponsored by 124 Senators and Congressmen of both political parties—would be a sign that Americans really do like children.

In the hearings Senator Mondale and I conducted on our proposal, witness after witness spoke of the crucial importance to later life of the earliest years.

Indeed, day care and in-home care; health, educational, nutritional and recreational services; prenatal and other medical care to reduce infant and maternal mortality and detect handicapping conditions; and a variety of parent and child support services in those critical early years are just what our bill would authorize.

The Child and Family Services Bill is not, of course, the first such effort for early childhood legislation. In 1971 President Nixon vetoed another measure Senator Mondale and I drafted, the Comprehensive Child Development Bill, to help provide for millions of preschoolers opportunities for the cognitive, psychological, nutritional and emotional growth essential to the lives of children.

Despite that veto, many of us in Congress have continued to work for such legislation.

The Child and Family Services Bill focuses on helping families obtain the services they need but cannot afford and the programs they want but cannot find.

Today the strains of unemployment and inflation have narrowed the choices for families in nearly every income group. Many women have to seek a second job and more and more mothers join the labor force every year.

Although many women prefer careers, most women with children work outside the home because their families need the money. The economic pressures are so great that millions of mothers work, undeterred by the fact of small children in the family. Estimates are that 51% of American mothers with children in school work outside the home; even one mother in three with a child under six has a job.

Not surprisingly, then, witnesses at our hearings reported that millions of American children suffer from poor health care, improper nutrition, and inadequate education.

This is the context in which, to reiterate, a number of us in Congress have been pressing the Child and Family Services Bill. It would provide a variety of vital services for parents and children and do so on a completely voluntary basis.

Is there a solid case for such legislation?

Here are some facts:

The infant mortality rate in the United States is higher than that of 13 other countries. Our bill would mean medical services to expectant mothers that would reduce the chance of infant death.

Some 200,000 American children annually suffer some handicap. Our bill would make possible early identification of handicaps and health care for new mothers and infants.

Six million children under the age of six have mothers who work but there are licensed day care centers for less than a fifth of these children. The Child and Family Services Bill would mean federal funds for part-day and full-day care for children as well as after school and summer programs.

There is then an obvious need for the kinds of services which the Child and Family Services Bill would provide.

That this need is widely recognized is indicated by the fact that the bill has been endorsed by such respected organizations as the AFL-CIO, League of Women Voters, American Bankers Association, United Auto Workers, United Methodist Church, Baptist and Lutheran Churches, National Parent Teachers Association, National Council of Jewish Women, and the National Council of Catholic Charities, to cite only a few.

Yet, as many readers of this book will be aware, last

year there was in many states in the country a deliberate effort to spread totally false information about the Child and Family Services Bill.

Unsigned, mimeographed flyers were—as I write, still are—being distributed by radical extremist groups alleging that the bill would let the government take over the up-bringing of children and make them wards of the state.

The allegations, outrageous falsehoods, have been re-peated in publications of the John Birch Society and similar groups and have caused an outpouring of letters to Sena-tors and Congressmen.

Fortunately, some of the nation's press has undertaken to respond sharply to the smear campaign against the Child and Family Services Bill, with publications ranging from the *South Bend Tribune, Michigan City News-Dispatch* and *Goshen News* in my own state of Indiana to the *Christian Science Monitor, U.S. News & World Report, UAW Washington Report* and *PTA Today,* vigorously denouncing the perpe-trators of the attack on the bill for their deliberate misrep-resentation of its contents.

The fact is that far from turning children over to the state, the Child and Family Services Bill contains at its outset the findings by Congress that "the family is the pri-mary and the most fundamental influence on children . . . " and that "child and family services programs must build upon and strengthen the role of the family and must be provided on a voluntary basis only. . . ."

Moreover, the proposal specifically prohibits any prac-tice which would "infringe upon or usurp the moral and legal responsibilities of parents or guardians."

Still further, the legislation provides that although the money for the programs will come from the Federal gov-ernment, they will be administered by state and local spon-sors. Each sponsor, moreover, must have an advisory council, which plays an important policy making role, made up of parents and persons recommended by parents.

As I write, in the spring of 1976, it is clear that opposition from the White House to the legislation coupled with the tight fiscal situation will prevent action on the Child and Family Services Bill by the 94th Congress.

Meanwhile, it is imperative to build a constituency in the United States of people who care about the needs of children and families.

This book, Volume II in the series, *Child Care: A Comprehensive Guide,* can play a significant part in developing the understanding necessary for that informed constituency.

The essays which Dr. Stevanne Auerbach has compiled in this second volume constitute an invaluable guide to shaping child care programs.

The book will help move us to the day when our great and wealthy country will make a genuine commitment to assuring that all America's children enjoy the decent health and education and opportunity which, 200 years after the birth of our nation, must be their birthright as well.

PREFACE TO VOLUME II

Setting up a comprehensive child-care program involves far more than decorating a cheerful room, locating a competent and loving teacher, and rounding up a group of neighborhood children. Romantic ideals inevitably clash with harsh realities.

The child-care movement not only incorporates the noblest intentions and ideals, but, unfortunately, operates in an area strewn with complex and exacting regulations, financial stumbling blocks, and a wide range of human problems involving children, parents, child-care workers, and the community.

The most common users of child-care services usually can pay only a fraction of the expenses involved, if anything at all. Therefore, child-care operators must turn to sources of government funding to establish and maintain their programs. In order to qualify for such funding, child-care programs must meet certain guidelines and standards established by various government agencies. In some cases hiring and personnel policies are dictated by a government funding source. Local building codes must be observed. As one seeks out information to enable compliance with the codes and regulations, one almost inevitably encounters confusing and even conflicting advice. All this adds up to months of painstaking groundwork before the first child can enroll in a program.

These are but a few of the number of complex issues that must be faced in the establishment of programs. The series will deal with many of these issues and suggest approaches that have been used successfully by the contributors.

This volume focuses on the planning of model programs and services. Not only do the brick-and-mortar issues have to be resolved, but decisions have to be made that determine the type and extent of services which are to be included in the child-care system. The examples included describe several of the programs that have often been discussed as model systems. These are usually communities where a great deal of energy, resources, and thoughtful planning have created effective child-care services. We will examine in greater detail what components have produced the results and trace the development of these models. The resources in the Appendix of this volume will refer you to additional models that have been described in other publications. For those who are interested in creating or expanding their own programs, the experiences shared by the authors will be most useful.

We begin the volume with the inevitable concern of financing. James A. Levine offers several strategies applicable to diverse geographical locations and programs. The references to helpful publications and organizations are aimed at helping child-care advocates with perhaps their most arduous and ceaseless problem—securing funding for their programs.

Programs started in Portland, Oregon; Appalachia; Berkeley, California; Denver, Colorado; and College Park, Maryland, all present useful models for viewing the range of complexity in planning and providing quality comprehensive services for each locale and replicating elsewhere.

Helen Gordon writes of her experiences as a primary organizer of the Portland, Oregon, community day-care program. The efforts in Portland won nationwide attention because of the remarkable success in attracting the active participation of a broad range of community interests. Gordon's report serves as a model for generating the community support crucial for the successful operation of a quality day-care program.

Jean Berman relates the history of the Appalachian Regional Commission's efforts to provide federally funded child-care services to a target area encompassing 13 states. Each of the states established its own agencies to coordinate and administer child-care programs, and within each state different systems for local administration were developed. The Appalachian child-care experience provides a look at the complexities and various methods of establishing state and local administrative networks for federally mandated and funded child care.

Kathleen Latham reports on the child-care project created by and for students and community at the University of Maryland in College Park. A way was envisioned to provide an ongoing service and linkage between the university, the public schools, and the community. The plans and development of the program are described, along with the ultimate frustration arising from lack of continued support for a successful model or demonstration program.

Kay Martin and Mary Millman recount the Berkeley, California, experience in organizing the community for expanded day care. In this account, the bitter conflicts that arose among competing interests in the community often severely jeopardized the program. The Berkeley experience points out the consequences of the failure of collective advocacy to remember the needs of all children, and the corresponding failure to measure progress by a concrete increase in service.

The last model is described by Mary W. Van Vlack, Ramon C. Blatt, and Paul T. Barnes, who discuss an approach to a systematic coordination and counseling service in conjunction with the University of Colorado Medical Center in Denver. The project provides a data bank of available child-care arrangements, support and training, and other kinds of information to families. The process and services could be adapted in other communities that want to bridge the gap between consumers' needs and community responsiveness.

In the process of examining the steps involved in a comprehensive view of child care, we have included history, politics, planning procedures, lobbying, parental needs, funding, and specifics for program components. These elements of care provide the additional supports and complete services important to the child and family.

Some succeed brilliantly and others fail dismally, but most child-care programs strive to complement the efforts of parents to provide their children with responsive, responsible, loving attention. In the traditional setting, the mother cares for her children at home all day and learns to recognize and meet their individual needs. But the typical child-care consumer lacks the resources and the time to attend to many of the family's personal needs. As much as he or she may regret it, the working parent often must choose between caring for the child directly or assuring the financial stability of the family. Time off from work jeopardizes employment; so the time left to spend with the child is brief. Consequently, the opportunities for sensitive appraisal of each child's subtle differences—and hence the responsibility for accounting for these differences—lies with the conscientious child-care provider.

With so many demands on their time and attention, working parents seek assurances that the basic health, nutritional, emotional, and developmental needs of their children are at least recognized by child-care providers. This volume contains papers on a wide variety of topics relating to these and other special needs of children in child-care programs. The authors address themselves to the need and means for providing services essential to comprehensive child-care programs.

Ann DeHuff Peters presents her proposal for health consultation teams as a means of providing health care to children in child-care centers who, because of their generally lower socioeconomic levels, often suffer from health care neglect. These teams also would provide mental-health care for the families of these children, and much-needed guidance for the parents, who often are unprepared for the stresses of parenthood.

Susan Aronson treats the specifics of the health consultant's role as a member of the support team. The consultant is faced with a variety of challenges depending on the particular program and its needs, resources, and outlook. As health support is critical to the effective comprehensive system, Aronson's recommendations and checklist are especially important for the awareness of both the educational and medical personnel.

Since many pediatricians advocate an expanding role for the nursing profession in the emerging specialty of nurse practitioners, we can understand more clearly the essential link nurses contribute to the availability of health services. Marilyn Chow describes her experiences in this role in a center which greatly benefited by the new form of outreach provided by a nurse practitioner.

The services, of course, once they are established, must continue to find a way to stabilize, expand, and redirect their impact. Almost every parent and staff person reports a need for more health education, information, and consistent care for the child, particularly if the child is mildly ill. All too often, we find the information provided by specialists does not reach those for whom it was intended. We hope that the recommendations for day care prepared by the Academy of Pediatrics will find a place in every center and home, and the benefits provided by these health specialists as described will be fully supported where the need for health care is greatest.

Health is just one of the critical components needed in comprehensive child care. Mental health services are another. The American Orthopsychiatric Association Study Group on Mental Health Aspects of Day Care presents a developmental point of view which requires consideration of those environmental and, especially, family influences that affect the child. Day-care services organized to contribute to mental health must consider the criteria used for the evaluation of the particular model of day care. Models for interrelating the family and day care based on the concept of the extended family and implemented by typical support systems used by families as resources must be channeled and supported by sensitive delivery systems focusing on children's needs.

Dorothy Nash Shack sees one important function of child-care centers as providing the opportunity to respond to early detection of problems and prescriptive services to promote the optimal development of children. Drawing on her experiences with the Early Growth Center in Berkeley, Shack explores the role of the psychologist as part of the team of the child-care staff and parents. She suggests ways to structure the child's environment and outlines a diagnostic procedure to determine each child's developmental needs.

David Brown calls upon his experience as a social worker with the Berkeley, California, Unified School District to outline the crucial role social workers should play in helping to assure that child-care programs supplement with sensitivity the family lives of the children enrolled in them. Communication between child-care providers and parents often suffers from a number of problems which the social worker can be uniquely prepared to alleviate.

Thus far, we have addressed health, mental health, and social services. Education and nutrition members of the comprehensive team are also included.

As the field of early childhood and education is resplendent in theory, practices, and programs, we will not duplicate here the many fine independent articles and materials readily available elsewhere. Many of these resources are cited at the conclusion of the volume. However, we felt the importance of including one major article that looks at the philosophy forming the basis of the educational program which affects the many other activities that will be included. If, for example, one sees the value of young children exploring their environment through a variety of shopping and food experiences, then the nutritionist would work closely in assisting the educational staff. Kitchens in child-care centers would be designed to be accessible and adaptable for children's use as a natural part of their educational program. Short trips to the local community stores would allow them the opportunity to have more concrete experiences. These are activities parents often engage in with children and are the times when so many important learnings occur. Through the understanding and the philosophy of Piaget, we can see many adaptations applicable to child care.

Keith Alward applies the teachings of Jean Piaget, perhaps the world's foremost authority on child development. Alward summarizes Piaget's theory of the stages of growth and provides specific illustrations of what to expect from children at succeeding stages in their development. From this, the developmental-educational component of child-care programs can be based on realistic goals for children, in the proper sequential order.

Linda Regele-Sinclair adds a rich experience in the area of food and nutrition as basic to the healthy growth of children. She contributes a menu of ideas vital to the fullest dimension of comprehensive services.

We hope that this second volume will put into perspective the gamut of the varied components which contribute to the whole of a comprehensive child-care program.

Volume III will explore the planning of creative environments in centers and homes and the development of infant care and family day-care homes. It will present approaches to the design and implementation of the places where children are, and indicate ways to maximize the experiences for all children.

Chapter One

HUSTLING RESOURCES FOR DAY CARE

James A. Levine

Most day-care programs, with budgets dependent on allocations from government bureaus, the community chest, and private philanthropies, suffer from the annual-grant syndrome. Characterized by frenzied proposal writing, speechmaking, and soliciting, the syndrome has plaguelike proportions: day-care directors all over the country scurry from meeting to meeting, red eyed and ragged from lack of sleep, trying to secure operating support.

Very little has been written about the hustle for funds and other resources, despite its importance to program survival; day-care directors are usually too busy writing proposals to write articles. However, day-care programs have developed a variety of imaginative techniques to enlist government and foundation grants, support from the business sector, and donated goods and services. It would be impossible to enumerate them all. Offered here are a few strategies which should apply to diverse geographical locations and programs, and references to several helpful publications and organizations.

WHERE YOU BEGIN: THE BUDGET

If a budget were merely a tally of the dollars and cents it costs to operate a program, the prospect of preparing one wouldn't be so mysterious and intimidating. Given limited resources, budget-making means not just juggling figures, but making difficult decisions about priorities and values: inclusion of one program component may mean exclusion of another; setting administrative salaries at a certain level may mean less money for staff members who work all day with the children; rewards to teachers for graduate credits may devalue the unschooled ability of others. It is precisely because a program's budget reflects so much that it is a useful tool for internal program analysis; it should always be available to parents, staff, and board.

Since the budget is also one succinct way of conveying to funding agencies how your program functions, the way you conceptualize and write it up can be important. Sometimes budget formats are prescribed by funding agencies. While you may have to comply with a prescribed form, you shouldn't adopt it as your own unless it is helpful in analyzing your program and presenting it to a wider audience. After all, you may seek funding from several agencies, and some will accept your way of doing things. So start with a format that makes sense to you.

Whatever format you choose, it should account not only for cash resources but for the cash-equivalent value of donated goods and volunteered services. Although these in-kind contributions represent between 5–25 percent of the resources in all good programs, they are often excluded from budgets. Only by including them can your budget reflect the totality of your program.

Helpful information on budget preparation can be found in:

Day Care: How to Plan, Develop, and Operate a Day Care Center by E. Belle Evans, *et al.* Boston: Beacon Press,

1971 (hardcover $6.95, paper $3.95). The best book to start with. Forty-two pages on "Planning a Budget" include two actual budgets—one high and one low—for a program of the same general standard of quality. However, the book does not deal with functional budgeting, a useful format now being widely adopted by state agencies and the United Way.

A Cost Analysis System for Day Care Programs by Eva C. Galambos, Ph.D. Atlanta: Southeastern Day Care Project, Southern Regional Education Board, 1971 ($1.00). This clearly written guide shows, step by step, how to prepare a functional budgeting system. The system differentiates and costs out as separate functions: management and administration, child care, food and eating, plant and maintenance, transportation, health, social services, and special functions. Available from the Day Care and Child Development Council of America, 1401 K St., N.W., Washington, D.C.

Manual on Organization, Financing, and Administration of Day Care Centers in New York City: For Community Groups, Their Lawyers, and Other Advisers, 2d Ed. by the Day Care Consultation Service, Bank Street College of Education, 1971 ($5.00). Although written specifically for New York City, the manual has wide applicability. A section on "Accounting and Management for Day Care Centers" outlines procedures that may help you monitor your income and expenditures. Available from the Bank Street College Bookstore, 610 W. 112 Street, New York, New York 10025.

A Day Care Guide for Administrators, Teachers, and Parents by Richard Ruopp, *et al.* Cambridge: MIT Press, 1973 (hardcover $10.00). Basically a condensation of Abt Associates 1971 report on program models and costs,

A Study in Child Care, this book should be a mind expander for program and budget conceptualization.

Day Care #7: Administration by Malcolm Host and Pearl B. Heller. Washington: Office of Child Development, 1971 ($1.25). One of several in a series of excellent handbooks on day care, this includes a chapter on "Decentralized Budget Development and Administration." Available from Superintendent of Documents, U.S. Government Printing Office, Washington, D.C. 20402. Ask for stock #1791-0161.

GOVERNMENT GRANTS

Although an array of federal programs supposedly support the provision of day-care services, actual support is not so easy to come by. Appropriations for day care have fallen way below congressional authorization, and a jumble of rulings to and from agencies administering funds has produced a situation in which dollars sputter inconsistently from the federal tap. Though funding at the state, county, and city level varies widely, it is often linked to federal funding patterns. Government funds at all levels seem to appear, like the smile of the Cheshire cat, without warning or duration.

Up-to-date information is crucial in dealing with the phantom funding situation. A few resources may be helpful:

Federal Funds for Day Care Projects (Women's Burcau Pamphlet 14, Revised), 1972 ($1.00). A listing of all sources of federal support, available from the Superintendent of Documents, U.S. Government Printing Office, Washington, D.C. 20402. Ask for stock number 2916-0010.

Day Care and Child Development Reports, published bi-weekly by Plus Publications, Inc. (2814 Pennsylvania Avenue, N.W., Washington, D.C. 20070), provides reliable current information on activities of the federal government that affect child care. If you can't afford the yearly subscription rate of $75, chip in with a few programs. If you can afford it, share copies with a less well-endowed program.

Your federal and state legislators—or their staff people —should have quick access to information about funding for child care. In seeking information from them, it's psychologically advantageous to be more than another voice on the telephone. Before you're in need of information, make an appointment to meet your elected legislator or the staff most likely to deal with child-care issues. You can use the initial appointment to ask for your legislator's opinions on child-care legislation and to explain your program.

It's not unusual to anticipate a call or visit from a representative of the government agency funding your program with feelings like those of an elementary-school child called to the principal's office: What did I do now? And so, some programs are content to leave well enough alone when the person assigned to monitor or advise them doesn't show up for months. Sometimes, however, the government-liaison person can offer useful critical commentary about your program, as well as information about funding trends. And if you make an effort to educate that person about your program, you may develop an ally willing to interpret your accomplishments to higher-ups and to keep you apprised of available funding. So rather than heaving a sigh of relief when your government agent calls to cancel an appointment, it might make sense to request a rescheduled one.

In some states and localities, administrators and elected officials are implacably hostile to day care, especially if it is being advocated by community or minority people. In such cases you may get help from representatives or field staff of nationally organized advocacy groups such as: the Black Child Development Institute, 1028 Connecticut Avenue, N.W., Washington, D.C. 20036; Children's Defense Fund, 1763 R Street, N.W., Washington, D.C.; Day Care and Child Development Council of America, 1012 14th Street, N.W., Washington, D.C. 20005.

Even with access to the appropriate literature and personnel, you may not learn about available government funds until shortly before proposals to secure them are due. Survival in day care often means working with a bureaucracy distinguished for its insistence on obtaining irrelevant information from program operators and its incompetence in providing important information, such as information about current funds available for day-care programs.

Case in point: One Friday at 4 P.M. a special delivery letter arrives at the day-care center announcing that the state department of community affairs—or any of a number of state bureaus—has $120,000 available for child care and is requesting proposals from programs across the state. All proposals must comply, of course, with the standard format consisting of seventeen parts, each with six subparts. The deadline for all proposals—noon the following Monday—leaves four-and-a-half official working hours for preparation of the center's proposal.

This situation and variants of it are no less outrageous than they are common. While boards, parents, and staff must protest these administrative nightmares and ward off their recurrence, in the short run they often have little choice but to cope. Here's one shortcut in the proposal writing game. Maintain one or more looseleaf notebooks with tabs corresponding to the standard proposal items: geographic area to be served, definition of need, services to be provided, history of agency, composition of board of directors, philosophy of program, educational curriculum, chart of personnel, résumés of personnel, budget, and so on. In the looseleaf you can retain photocopies of pertinent information from previous proposals, typed without headings or roman numerals. Each time you have to do a new proposal, fill in the headings and roman numerals as necessary. Even if you have to do substantial rewriting, you'll save many precious hours and a good deal of emotional drain.

FOUNDATION GRANTS

While the Comprehensive Child Development Bill was supposedly en route to passage, several of the major national foundations supported a limited number of day-care programs, technical assistance groups, and new day-care publications. Theory was that foundation capital invested in day-care projects would demonstrate how the inevitable federal dollars could most effectively be spent. With President Nixon's 1971 veto of the child-development legislation and subsequent federal cutbacks in social services, national foundations have been besieged with requests for operating support from day-care programs. The simple fact is that the major foundations can't possibly take on the job that should be done by the government, and they are reluctant to fund any program that isn't highly innovative, replicable in other settings, and—ironically—likely to encourage governmental support to ensure its continuation. If you need money to operate your program or to develop a new program component, try your local or regional foundations. Money is tight everywhere, but at least the local foundations are in a better position to understand the importance of your program to the community.

There are several ways to find out about foundations, their areas of interest, and their grant-making policies:

The Foundation Center (888 Seventh Avenue, New York, N.Y. 10019; 1001 Connecticut Avenue, N.W., Washington, D.C. 20036). While it does not direct applicants for funds to particular foundations, the Foundation Center is *the* library for seeking information about foundations, with current files on the activities and program interests of over 30,000 foundations in the United States. The center's ever helpful librarians can provide you with:

A computer print-out listing all foundation grants over $5,000 made since 1971 in any program area. A request for all grants made in day care, child care, and early education will pretty well cover the field. The fee for this service is $15 for a minimal print-out of 50 grants and 20¢ for each additional grant listed. To receive a print-out you must submit a search-request form, available from the center or any of its regional collections. The center will soon have listings of all grants made for day care printed on microfiche, making this information available more quickly and less expensively.

A free booklet providing advice on *How to Find Information on the Foundations.*

Referral to the center's regional collections (Atlanta, Austin, Berkeley, Boston, Chicago, Cleveland, Los Angeles, St. Louis) and to other sources of information about foundations.

Taft Information System. The most efficiently organized guide to the major foundations includes brief descriptions of their programs, guidelines for applications, pertinent information about individual members of the board of directors, and a list of recent grants. Because the *Taft Information System* is very expensive, it is not widely available. Your library is not likely to own it, but the development office of a local college might; or contact Taft Information Service, 1000 Vermont Avenue, N.W., Washington,D.C.

The Foundation Directory. Provides information about the purposes, finances, and officers of over 5,000 foundations with assets of $500,000 or more or annual giving of $25,000 or more. Entries are updated in the Foundation Center's "Information Quarterly." Though not as informative about individual foundations as the *Taft Information System,* the *Directory* can do what *Taft* can't: steer you to your local foundations. And it should be available in the library or in the office of any agency that depends on foundation grants.

Foundation News. This bimonthly publication of the Council on Foundations (888 Seventh Avenue, New York, N.Y. 10019) includes a section listing, by state, grants of over $5,000 made during the previous two months. It also features articles about individual foundations, the interplay between government policies and foundation giving, etc.

Of course, there's a long way between knowing about the grants a foundation has made and receiving one yourself. The best way to approach a foundation is with a letter or brief proposal outlining clearly and succinctly why your program is needed, what it will do, how it will do it, who will do it, how much it will cost, and how long it will take. Inflated rhetoric about the needs of your community or your ideals is likely to carry little weight, though, because no matter how sympathetic foundation officers are with your goals, they need to know what differentiates your program from others and to glean some idea of the specific results to be produced by an investment in your work. For more detailed advice on proposal writing, see:

> "What Makes a Good Proposal?" by F. Lee and Barbara L. Jacquette, *Foundation News,* Jan./Feb. 1973, pp. 18–21. Written by the treasurer of the Carnegie Corporation and an officer of the Foundation for Child Development, this invaluable article outlines the features of a good proposal and some of the criteria used by foundations in assessing proposals.

> *The Bread Game: The Realities of Foundation Fund Raising,* $2.95. A paperback available at some bookstores or from Glide Publications, 330 Ellis Street, San Francisco 94102. This simple and step-by-step guide is little known outside the San Francisco Bay Area, but enormously helpful.

VOLUNTEERS

With government funding so precarious and foundation funding so difficult to obtain, very few day-care programs survive without substantial reliance on contributions in the form of goods or volunteer services. At one program, for example, a certified public accountant from a top-level firm helps manage the books, the pledges of a college fraternity helped construct the playgound equipment, a student from a dental college helped develop a program of dental-hygiene education in family day-care homes, a secondary-school student is shooting videotape which will later be used for in-service training, a psychiatric social worker from a local clinic donates one hour per week for staff training, and a National Guard unit will soon be installing a heavy wire fence around the playground.

It would require a separate article to suggest all the ways in which volunteers contribute to day-care programs. The chart included in the Appendix of this volume gives a quick rundown by specific area of program need. The only constraints on soliciting volunteer help seem to be the human imagination (the yellow pages are a good mind-expanding guide) and fear of asking: it's simplistic, but true, that people like to feel needed and be appreciated for their unique ability to contribute.

Since personnel costs usually represent from 70–80 percent of the typical day-care budget, it may be worthwhile to give special attention to one increasingly available pool of volunteers: students from high schools and colleges, more and more of which are granting academic credit for field work and internships. High school and college interns can be a tremendous asset to any staff. They are eager to get into the "real world" and often go all out to help. They tend, however, to have the curious habits of getting sick around exam time, and of having prior commitments around school vacation time. They are often unaware of how their convenient reversion to student status and the academic timetable can leave a program in the lurch and anger regular staff, who have no such pause.

One mechanism for promoting mutual satisfaction between day-care staff and interns is the internship contract, which is, in effect, a letter from the day-care director and staff who will work with the intern, co-signed by the intern and the intern's faculty sponsor. The letter should outline job expectations as specifically as possible, including whether or not the intern is expected to work during school vacations. If the job is to change, or be fluid, the letter should say so. The letter can indicate when the sponsor will be able to observe, and how and when the intern or the agency will provide an evaluation. And, most importantly, the letter can allow the intern and the agency a two-week trial period at the end of which either can cancel the relationship with no explanations, no hard feelings, and no jeopardizing of future internships.

Though the intern's contract does not guarantee a good working relationship, it is no mere verbiage or piece of petty bureaucracy. There is a psychological bond implicit in a signed contract between an intern and his or her future colleagues at the center; it helps make the intern feel like part of the group and impresses the importance of his reliable contribution to the functioning of the center. It clarifies to all staff members just what the intern is expected to do and therefore helps them in guiding him. And, by making explicit the mutually supportive relationships necessary in day care, it increases the likelihood of a successful internship.

Other material pertinent to work with interns is available from the National Commission on Resources for Youth (36 West 44th Street, New York, N.Y. 10036), which offers technical assistance, curriculum materials, and a model for involving teen-agers in the day-care center in a systematic way, and the Education Development Center, Inc. (15 Mifflin Place, Cambridge, Massachusetts 01238), whose publications in the Exploring Childhood Program contains a variety of exercises that can be adapted and used effectively to orient and train high-school students.

INDUSTRY AND DAY CARE

Industry and day care; juxtapose these two words and people automatically conjure up an image of a day-care facility attached to and operated by a factory. But industry-based day care is just one way for the private sector to support day care. Mary Rowe, addressing the Second National Conference on Industry and Day Care (1972), outlined 17 ways in which business and unions can become involved in child care, including: donation of money, services, or space to a community child-care program; donation of matching money under Title IV-A; paying the salary necessary for a company-based child-care service which would help employees find day care for their children; underwriting day-care research; lobbying for day-care legislation; and developing a training and licensing program for family day-care homes.* In addition, industry can support child care by adopting more flexible policies on working hours. Half- and three-quarter-time positions would allow more men and women to combine working and parenting in a healthy way.

What arguments can be made for industrial support of day care? Unfortunately, there is little hard evidence to support the contention that employer-subsidized child care increases the profit margin. The findings of *Employer-Subsidized Child Care*, a 1972 report to HEW prepared by Inner-City Fund management consultant Donald Ogilvie, suggest that "except under unusual circumstances, savings from employee turnover and absenteeism are likely to be relatively small; on the other hand, the recruiting value of subsidized day care may be appreciable." However, the first annual report (1972) of the Northside Child Development Center in Minneapolis, a program supported significantly by several local industries, indicates a 21.4 percent reduction in absenteeism among mothers with children, a substantially lower rate of turnover for mothers using the center compared to those in comparable jobs who do not use the center, and an increase in job performance and satisfaction (an issue which the Ogilvie report admittedly did not address). If Northside's results are maintained and the data well publicized, they may help induce other indus-

*All 60 pages of the Conference proceedings are available from Urban Research Corporation, 5464 South Shore Drive, Chicago, Illinois 60615. Though they include valuable information on industry-related day care, they cost $15.

tries to support child care. In the meantime the industrial sector, which claims a desire to be socially responsible, must be educated about the ways in which business policies and the availability of child care might interact to support family life.

On Hustling and Educating

Correcting our nation's inadequate and fragmented funding of day care will require nothing less than a sustained political fight and a massive campaign of grass-roots, highly personalized public education. For although day care has been much publicized in recent years, a surprising number of people—people whose opinions influence the availability of local, state, and national funding of day care—don't really know what day care is. Many of these people— elected officials, business persons, bankers, midlevel bureaucrats in city and state government, newspaper editors —would be willing to learn about day care and even to visit a center, but they have very busy schedules, and they've never been invited.

One way to begin raising the consciousness of these people is to invite them to have lunch, with the children, at the day-care center (or in a family day-care home). Spilled milk, clanking dinnerware, and laughter may not, at first thought, seem to provide the proper ambiance for serious discussion of your program; and lunch probably isn't the best time to convey what the kids do at the center. But lunch with the children at the center is a convenient, if unorthodox, adaptation of the business lunch, to which many of these people are accustomed. It will give your visitors a feel for the place and will foster a spontaneous intimacy that is quite difficult to achieve in more business-like settings or during much-labored-over community education nights, rarely attended by people not already involved in day care. It will give your staff a chance to talk about the program, answer questions, and remove some of the prejudicial stereotypes people have about day care. Lunch at the center can't help but increase community understanding of your program. Some of your guests may want to extend their stay or return for a morning or afternoon visit. And if your program publishes a newsletter, you can add the names of your visitors to your mailing list and thereby provide them with a gentle monthly reminder of their visit.

Neither lunches nor newsletter will guarantee successful fund raising. In fact, if you play the part of an aggressive salesman rather than a host and friend, your efforts will probably backfire. But a program of public education geared to individuals and mounted at the grass-roots level might, in the long run, have significant impact on public attitudes about day care. In the short run, such efforts are crucial in local day-care politics: if you're going before the city council to seek revenue-sharing funds for day care, it can't hurt if three city councilors have lunched at your center. And public education may have serendipitous effects: because a local dentist has heard, via the grapevine, that you are trying to do some community enlightening, he may be more willing to help you develop a dental screening program; or one of your lunchtime guests might be appointed a trustee of a local foundation.

It may be that the best way to hustle is when you're not hustling, but educating!

Chapter Two

A COMMUNITY ORGANIZES FOR CHILD CARE: THE PORTLAND EXPERIENCE

Helen L. Gordon

Beginning in January, 1967, planning began in the Portland, Oregon, area to provide child care for the thousands of low-income families in which the mother was prevented from seeking employment by the demands made on her by her family responsibilities. The Portland experience, in which I played an active role, has many lessons of use to almost any community attempting to initiate a child-care program.

The first step was the formation of an ad hoc committee on day care, set up by the Volunteers of America under the sponsorship of the Portland Metropolitan Steering Committee, the Community Action Agency in the area. To insure that the committee represent a broad range of potential users and providers of day- and night-care services, we made a special effort to include a diverse cross section of the community on the committee. Toward this end we reached wide-ranging definitions of "user" and "provider."

24

"Users" were not just families needing care for their children, but also business and industries needing female employees; public schools desiring motivated and well-prepared children; high schools, community colleges, colleges, and universities wishing to relieve some of the burdens of motherhood from their students; law enforcement agencies wishing to cut down the incidence of juvenile delinquency so prevalent among "latch-key independent kids"; and employment agencies needing substantial pools of female workers.

"Providers" included day-care providers, public, and proprietary; business and industries interested in providing directly for the child-care needs of their employees; public schools, community colleges, colleges, and universities, which might invest some of their buildings, equipment, and manpower for such services; churches, which might contribute the use of their buildings and nursery school equipment; civic and service groups, which might contribute funds in an effort to improve the well-being of their communities; and foundations, businesses, and trust funds, which might contribute financially as an investment in the health of the community.

Early in 1967 I was asked by the regional Office of Economic Opportunity office to leave my job as director of the preschool at the Jewish Community Center and join the local community action program staff (Portland Metropolitan Steering Committee) to begin planning for child-care services.

The first meeting to organize the committee, which I chaired, was held in January, 1967, and had 47 people present. By mid-February 97 persons were willing to serve, and by early March the number had increased to 129. We created seven subcommittes: structure, staff needs and training, infants and toddlers (family day and night care), preschool group care, school age, emergency services, and health services.

Representatives from the following groups served on these subcommittees:

Labor. Officially, Oregon Bureau of Labor; also, several labor unions with large female memberships

Civic and service groups. Junior League; Council of Jewish Women; American Association of University Women; League of Women Voters; Council of Churches

Business and Industry. Pacific Northwest Bell Telephone and Tektronix

Community agencies. Departments of welfare and health, public schools (including Head Start), employment and concentrated employment program, Vocational Rehabilitation, family counseling, Portland State University, Tri-County Community Council, Portland Metropolitan Steering Committee, and subsidiaries from neighborhoods

Day-care providers. United Good Neighbors (UGN) and UGN day-care agencies; proprietary day-care programs

Political sector. County commissioners, city councilmen, state legislators, state office of Economic Opportunity.

By May, 1967, the subcommittees had completed their task, and nine people, brought together as volunteers or paid by the Urban Studies Center, Portland State University, through Title I Higher Education Funds, worked for 13 hours to put all of the minutes and subcommittee findings into one report called *DUCS* (Day Care Urban Coordinating Services).

This report and the entire process received national attention as a remarkable example of diverse segments of the community working together and contributing to a vital local program. In addition, the poor gained access to the policy-making process and contributed to the formulation of new patterns for the delivery of services.

In addition to the written report, the committee entered into verbal agreements in a number of areas. Churches agreed to provide space. Schools offered to sponsor and operate day-care programs for their pupils, using the resources of other community agencies. The Housing Authority of Portland agreed to investigate the possibilities of providing space. Schools and health departments—including mental health, welfare and family counseling—agreed, if funds could be located, to provide specialized skilled staff, such as social workers, nurses, doctors, psychologists, child development-supervisors, family life educators, and nutritionists, to various child-care programs.

The committee agreed that it intended not to create new agencies but rather to expand and upgrade the services of existing agencies, public, private, and voluntary. To accomplish this, the committee undertook to develop coordination at local levels through neighborhood points of intake, information, and referral (DUCS). Families would use these neighborhood centers to assess their needs and locate the services offered.

The Portland Metropolitan Steering Committee formed the nucleus of a larger committee established in 1968 with the beginning of Model Cities planning in Portland. Simultaneously, planning began for the Parent-Child Center Project, from which the governor of Oregon convened a conference of the public and private sectors, focused on children and emphasizing those under the age of three from poverty families.

In the beginning of 1969 the federal government asked Portland to become a pilot community for Community Coordinated Child Care (4-C's). The metropolitan area 4-C's council, encompassing three area counties, drew membership from a broad spectrum of community organizations including:

The Political Sector. City, county, and former state legislators, and the governor's assistant on human resources

Public Agencies. Departments of welfare and health, public schools, and Head Start, community colleges (three), Portland State University, Division of Continuing Education, employment and concentrated employment program, Bureau of Labor, mental health, and the governor's committee on Manpower

Civic and Service Groups. Council of Churches, National Council of Jewish Women, Junior League, League of Women Voters, American Association of University Women, Portland Metropolitan Steering Committee, the county community action programs of each of the participating counties, Head Start, United Good Neighbors, and Community Council

Business, industry, and labor. National Business Alliance, Pacific Northwest Bell Telephone Company, Jantzen Knitting Mills, Pendleton Woolen Mills, Kaiser Hospital, Amalgamated Clothing Workers of America, and Communication Workers of America

Parents. Elected by parent groups and comprising one-third of the board

Proprietary day care. Represented by individual operators and the Oregon Association of Day Care Operators, Portland Area

Voluntary day-care agencies. Volunteers of America, St. Vincent de Paul Society, and other day care nursery schools.

The 4-C's council consists of a board of 23 and an executive committee with the same ratio of parents. In fact, a low-income parent serves as president of the council. The council staff consists of an executive director, a business manager, and a secretary. The staff needs grew to an assistant director, another secretary, and an assistant bookkeeper.

The council has ten committees, each chaired by a member of the board of directors and including members from all of 4-C's and some from outside. Several of the committees are described below:

Community resources. Responsible for securing local funds and other resources; public information and education and public relations

Community sharing. Establishes joint activities among day-care providers in areas of training, transportation, parent activities, exchange of program materials, joint purchasing, etc.

Finance. Works with 4-C's business manager in relation to operating and subcontracting budgets

Legislation. Works with other groups, such as the Day Care and Child Development Council of Oregon, in preparing legislation for the state legislature, and in doing digests of Federal legislation

Manpower. Assesses future needs, recruitment, and training of staff and development of career ladders

Program review. Monitors and provides technical assistance and consultation for operation day- and night-care services for which 4-C's has contracted.

The future goals of the council include attracting the membership of an even broader cross section of the community, including other civic and services groups, businesses, industries, and labor groups and securing a larger investment of funds from business and industry. We also continued and expanded one of our most important activities—informing various government agencies and bureaus of state and Federal statutes which can provide funding for various programs. The council serves a valuable function in providing a channel of communication for the various government agencies able to deal with child-care issues, and also in researching and lobbying among public policy makers.

After the plan for day-and night-care services was submitted to and approved by the regional federal office on July 28, 1970, public welfare signed an official contract with the 4-C's council. The contract provided that locally raised funds from private sources and public in-kind support be matched by state public welfare. Also under the contract, the 4-C's council is permitted to subcontract with provider agencies.

Subcontracts for the operation of day-care programs were signed with a variety of providers. Colleges and universities in the area, and even a high school vocational training course, have established day-care centers as training labs. Various industries have established centers for the children of their employees. The public schools have established after-school programs for children of working mothers who otherwise would return unsupervised to empty homes. The Council of Churches helped to establish many centers in church facilities.

Other contracts were prepared with such groups as providers of transportation for the children, the Home Extension Service to supply family-life educators and nutritionists, county health departments for social workers, social work aides, and nurses, and with the Portland public schools to supply child development supervisors.

The 4-C's council continues to seek support from a number of federal agencies, including the administration on aging, law enforcement (juvenile delinquency prevention related to school-age day care), the Office of Economic Opportunity, the Office of Child Development, the Social and Rehabilitation Administration, Social Welfare and Training, and the Department of Labor.

Many lessons applicable to any community desiring to establish child-care programs have come out of the Portland experience. Without broadly based community support and involvement, the many existing resources of the community cannot be tapped to contribute to the total child-care effort. One thing we learned was that the entire child-care program affects every segment of the community, and every segment can make a substantial contribution to solving the problem. Dedicated, hard-working leadership can perform a crucial service in informing the community both of the nature of the problems and the role everyone can play in solving them.

Chapter Three

ORGANIZING SERVICES FOR APPALACHIA'S CHILDREN

Jean H. Berman

Appalachia stretches along both sides of an ancient mountain range that takes in thirteen states from southern New York to northeastern Mississippi. The mountains are austere and beautiful; the people are proud and independent. Living is hard. Until recently the economic indicators of the region have pointed up poverty and isolation—limited access to the jobs and services essential to economic growth and personal well-being. Now the region is beginning to change. New highways, hospitals, and schools have been built, and with them have come health services, manpower training programs, and child-development centers.

The changes started in 1965 when Congress established the Appalachian Regional Commission (ARC) to "develop, on a continuing basis, comprehensive and coordinated plans and programs . . . to assist in development of the Appalachian Region."* The highways came first, to

*Appalachian Regional Development Act of 1965, as amended, Section 102.

alleviate the isolation. Brick-and-mortar programs provided the needed facilities from which education training and medical services could be delivered to develop the region's human resources and establish the basis for economic development. In 1969 Congress added to the ARC program by approving funds for programs in nutrition and child care. Planning to provide for the full development of its most basic resource—the region's children—began early in 1970.

THE NEED

As in the rest of the country, the delivery system in Appalachia has been fragmented. Over 200 federal and state programs relate to children, but there has been no way that any but the most knowledgeable and sophisticated professional could guide parents to the range of needed services for their children. In Appalachia the problem has been compounded; many services simply have not existed, and those that do often have been inaccessible. Not only were services needed, but new ways of delivery had to be devised.

The overriding goal in planning investments in prenatal care, in infant development, and in early childhood up to the age of six was to assure an array of preventive services through establishment of an organizational framework for the delivery of those services.

Planning is the key to organization of a sensible service system. Because the major sources of service for children are state agencies, and because the bulk of federal funds flow through state agencies, it was unrealistic either to plan for just part of a state, or to have primary planning done by agencies other than the ones directly responsible for providing service. The problem is that agencies traditionally plan for their own activities. Since child development is not the "property" of any one state agency, an effective coordinated program requires that all plan together. Comprehensive services require state action and may require adjustment of state regulations. Thus the process of organizing services began with coordinated planning at the state level.

THE FRAMEWORK

The conceptual framework for planning was developed by the Appalachian Regional Commission:

1. Conservation and development of our human resources must start before the child is born and continue during his youngest years.

2. The needs of infants and very young children are many and varied.

3. "Child development" encompasses a far broader range of services than those covered by categorical health, education, or welfare programs.

4. Services should be comprehensive and should be of a variety and quality to promote optimum development.

5. States must do their own planning so that individual state and local needs are prioritized appropriately.

6. Planning for service is intimately involved with developing a system for administrating the program.

States then began to build on that basic framework.

STATE PLANNING

Appalachian governors issued executive orders determining that each agency responsible for services to children should be represented on an interagency child development committee. At the very minimum, health, education, mental health, and social services are represented. Also included is the governor's Office of Planning. Some governors included Community Coordinated Child Care (4–C) committees, private agencies, and consumers as part of, or advisory to, the state planning group.

A further step away from the categorical approach led to the decision that planning would not be housed in any of these agencies, but would be a separate entity in the governor's office, the state's planning office, or other suitable agencies. The actual planning would be done by the interagency group and the staff it employed.

The committee was charged with developing a statewide plan for the coordinated delivery of services to children. Such a plan would "devise ways of maximizing the effectiveness of existing resources, pinpoint needs and priorities and assure interdepartmental cooperation"(1).

The planning process was undertaken systematically, starting with identification of each of the components of a comprehensive child-development program. Health (including nutrition), education, mental health, and social services were broken down into a clearly defined array of preventive, diagnostic, and treatment services. Usually each agency represented on the planning committee was responsible for cataloguing the services that each agency currently delivers as well as projected or needed services. This catalogue of services was the first step in planning. Quality criteria and performance standards were attached to each component.

Concurrent with this activity was the collection of demographic data such as numbers and location of children, socioeconomic status of families, mortality morbidity statistics, to name a few of the most important items. Analysis of this information combined with a geographic determination of existing services provides the baseline for needs assessment.

Perhaps the most important part of needs assessment is asking the people what they need. Thus, a major link in the planning process was the establishment of substate planning groups.

Substate Regional Planning

Because of its low population density and the paucity of resources, Appalachia has developed multicounty planning and development districts as a mechanism for economic and human resources program development. These organizations, whose boards of directors are made up of at least 50 per cent locally elected officials, are called Local Development Districts (LDD's). In most of Appalachia, LDD's became the base for local child development planning. No new competitive agency was established. Planning would utilize existing structures. Subcommittees to LDD boards were established to assure appropriate local input. Usually those committees utilize the 4–C's membership criteria of one-third consumers, one-third public interests, and one-third private interests.

The first task of these substate planning groups was to utilize the components, performance standards, and demographic data from the state planning effort to survey and assess local resources. Identifying local resources is often a difficult task. In most communities, services may come from many sources other than state agencies. Private physicians, proprietary day-care centers, churches, and voluntary health and welfare agencies are only a few. Resources were identified; county and multicounty profiles were developed.

Next steps were more complex, for local planning groups had to develop a list of program priorities—what services were needed most, how were they to be delivered, were new facilities required, what costs were involved?

By necessity, local planning was carried out within political and economic realities; seed money was available from ARC. Federal sources specified service to a limited group within the low income category. That knowledge affected priority setting. What services would states, local governments, and consumers be willing or able to support when federal funds were so limited? That question was answered slightly differently in each planning jurisdiction. Some communities set the establishment of comprehensive services in a group day-care setting as a priority; some designed maternal and infant health-care services; some expanded existing nutrition programs; some developed outreach delivery mechanisms for parent and child education; two states designed emergency medical service for newborns.

PROGRAM ADMINISTRATION

The purpose of all this planning was, of course, to provide better health, education, nutrition, and social opportunities to a target population of Appalachian children. Planning pays off only when more effective and efficient services are delivered; when there is an opportunity to assure a child's growth and development; when parents can see the need for and then utilize preventive services.

But behind this goal was the need to begin the process of developing a way for families to tie into the range of services they might want or need for their children. Bureaucratic language terms this "development of a single entry to a system of comprehensive services." Design of a single-entry system requires both coordination with and cooperation of the variety of service providers in a given community or county and the development of an administrative mechanism to implement the plan. Thus, the planning mechanism established program objectives, service components, and policies by which projects could be implemented and assessed. Overall administration and management of the variety of services was the next step toward service delivery at the local level. Decisions need to be made at each level of program information and must flow in both directions; thus coordination is essential if planning is to result in service programs that make a difference.

Child development planning in Appalachia has resulted in several administrative models of coordinated service delivery. In West Virginia, the governor has established through executive orders a state Office of Child Development within his own office. In North Carolina, the Office of Child Development is located in the state Department of Administration. Ongoing program monitoring, carrying out of planning policy, and technical assistance to regional administration units is the role of that office. The state Department of Community Development in coordination with the Department of Social Services is the overseer of Ohio's Appalachian Child Development Program.

Location of the state level administration directly in the governor's office or in his administrative department has several advantages. Clearly, it demonstrates the authority that this activity carries. In addition, administration is free of the categorical priorities that might be imposed by placement in a service agency. Placement in a line agency may reduce the cooperation of other service-providing agencies.

In New York and South Carolina, Appalachian programs now emanate from state departments of social services. In each of these states, location in an operating line agency is the logical "spinoff" from their original location in the governor's office during the planning phase of the program. Where operational responsibility is a major focus, activity can be carried out best in an operational agency. Contracting for service plus monitoring and evaluation are the major tasks.

State policy is the skeleton around which programs are built. State policies deal with overall issues such as priorities for type of service and target populations, the quality of service, the nature and comprehensiveness of a service, participation of those establishing and/or utilizing services, local financial participation, coordination, and linkages.

Although the structure of an administrative mechanism for program operation is essential at the state level, the key to coordination and service delivery for rural areas has been the local development districts. For it is at this level that policies from the state and priorities from local groups evolve into operational programs.

Multicounty planning and development districts then become the focus where the skeleton is fleshed out and the policy evolves into actual programs for rural families. Program operation is coordinated here.

In much of rural Appalachia it is sensible to join counties together under the umbrella of the Local Development District to centralize administration, training, specialized consultation, community development and public relations, and monitoring. Centralizing these functions allows freedom at the local level for the delivery of service and assures program accountability to the state.

The decision as to what services shall be developed first—medical and dental services, nutrition, infant stimulation—was part of the planning. Defining the delivery system (child-care centers, home outreach workers, stationary clinics, mobile clinics, etc.), and establishing intake and referral mechanism (intake at a central point, intake from a number of community agencies, intake through home outreach) are administrative tasks.

At the community level, local people provide services to their neighbors who need or want them with consistent support from the district level. Direct service to children and families on a personal level is the major task at the service delivery level.

This discussion has focused on the development of programs from the state level downward to the local level. The system works efficiently only when there is response to the needs, problems, and progress of the local level. This mechanism provides for open information flow from the local level to the multicounty district level to the state level where ongoing planning and policy development can assure ongoing effective service delivery.

STAFFING

Little has been said about staffing at each program level. Appropriate staffing is the key to the successful operation of a planning and administrative system for the delivery of comprehensive child-development services. Just as "child development" encompasses a range of categorical programs, so the professionals who plan and administer these programs must be able to bridge the traditional areas of administration and planning. Appalachia is developing this new type of professional competence that calls on skills such as systems design, health planning, early childhood education, economic development, community organization, public relations, accounting, and political science. Rarely are all these skills found in one administrator. As a result, staffs are made up of individuals with these competencies who themselves are a coordinated body to plan and administer programs. Certain skills are appropriate to state level activity; others are present in the local development district. All skills are available to any unit of the program.

Appalachian Achievements

In Appalachia, signs of progress are visible everywhere. These signs can be seen at many levels concomitant with the growth of Appalachian Child Development planning and program development. The states of Georgia, North Carolina, and Kentucky have established state administrative departments of human resources, bringing together the previously separate category agencies that service human needs—health, social services, manpower development, commissions on youth, the aging, and others. Tennessee has established an Office of Child Development to assure coordinated planning of all children's services.

Local Development Districts are assuring that child-development programs are coordinated with development of a broad range of human resources and economic development plans. In addition, scarce resources are brought together so that duplication is avoided and service gaps are filled.

Services are targeting in on children and families. Statistics show a decrease in infant mortality where new projects provide prenatal care and infant emergency medical services. Screening programs have led to prevention or early amelioration of handicapping conditions; children are receiving quality day care where none previously existed. And throughout Appalachia there is a new awareness that communities can assist parents in providing a better quality of life for children.

Note

1. Lazar, Irving. "Organizing Child Development Programs." Appalachia 3 (1970); Washington D.C., Appalachian Regional Commission.

Chapter Four

ORGANIZING COLLEGE CHILD CARE

Kathleen B. Latham

During the spring of 1972 students at the University of Maryland, in College Park, Maryland, felt the need for day care with a strong educational component. Several unsuccessful attempts were made to seek university funds for a child-care service for university students, staff, and faculty. The university administration consistently rejected offering any financial support to such a project. But it was discovered that state funds were available under Title III of the Elementary and Secondary Educational Act of 1965 (ESEA). These funds are specifically designated as "grants to local educational agencies for innovative and exemplary programs to demonstrate ways of making a substantial contribution to the solution of critical educational needs."

Over a period of months these students, in their quest for an educational child-care service at the university, took the initiative to seek support from interested faculty members, the Chancellor, the Governor, and several key legislators in the Maryland House of Delegates. During this process it was found that Title III funds were granted to projects which must function through a local public education system. Therefore, support was sought from the staff of the board of education in Prince George's County, where the University of Maryland is located.

University faculty, students, and the staff of Prince George's County Board of Education planned together to write a joint Title III proposal to be submitted by the board of education to the Maryland State Department of Education for approval and funding. The grant, therefore, would be made in the form of a contract between the Maryland State Department of Education and the Prince George's County Board of Education.

It was suggested that part of the proposal should include the development of a "rural center" component. The child advocacy program (also under supervision of the Prince George's County Board of Education) acted as a catalyst to the development of a day-care center in the rural parts of the county. Meetings which were held with the local Child Advocacy Council and local community representatives clearly spelled out the local interest, need, and support of this kind of activity in the Baden and Brandywine area of Prince George's County.

Dr. James A. Sensenbaugh, the Maryland state superintendent of schools, stated well the necessity for interagency cooperation for developing programs for young children: "We in education believe that no one agency alone has the resources to meet the multiple needs represented within programs for young children. Therefore, it becomes necessary for all agencies to work together coordinating program planning, implementation, evaluation, and resources." (1)

The possibility for interagency cooperation became apparent. Contacts with staff members of the Department of Social Services were initiated at county and state levels.

Additional meetings took place to identify and clarify the roles and responsibilities of all agencies involved.

The Office of Education responded with enthusiastic support and technical assistance in the development of the proposal.

Planning the proposal of this project involved input from many sources. From the time that the Prince George's County School Board approved the preliminary design, it provided a great deal of technical assistance. In addition, the University of Maryland directed the College of Education to participate. Professors and staff enthusiastically supported this project and made many valuable contributions.

Funds ($180,000) for the project, which was entitled "Growing Together," were awarded by the Maryland State Department of Education, who considered the proposal one of the top priorities in the field of education in the whole state of Maryland. Federal funding of the project was to be supplemented by fees based on a sliding scale of parents' ability to pay (that is, income and the number of dependents).

Two early childhood development centers were designed—one center to serve the University of Maryland staff, students, and faculty, and the other center to serve the rural community of Brandywine and Baden, Maryland. Due to problems in locating space in the public schools, the university area was split up into two separate centers. Gaywood Elementary School consisted of five classrooms and the main office of the whole project and served approximately 70 families. Greenbelt Elementary Center consisted of two classrooms serving approximately 30 families.

The "rural center" component of the project was located in the southern part of the county in Brandywine and Baden serving approximately 40 families. The Baden center was designed to cooperate with the already existing South County Day Care Center operated by the county Department of Social Services.

For a thorough breakdown of the centers and the number of staff members and children per center, a proposal of staff and group composition is included which was projected for the second year of the project's operation.

Upon the initial funding of project Growing Together, a steering committee was formed, composed of representatives from the four main agencies involved in the interagency effort—the University of Maryland in College Park, Prince George's County Department of Social Services, Prince George's County Board of Education, and Prince George's County Child Advocacy Council. These four agencies worked together in setting up and advising the project's initial operation and in interviewing and hiring a project administrator.

Alice Clark, the project administrator, then helped the committee choose two project coordinators (designated by the county board of education's regulations on job titles as "helping teachers"). These three people, comprising the administrative staff of the project, with the help and guidance of the steering committee then began to implement the beginning stages of the project—interviewing and hiring a staff, ordering the equipment and supplies, designing enrollment procedures, and developing a two-week training-orientation workshop for the incoming staff.

After six months of successful operation, it was learned that the county's local board of education was considering a vote on a resolution not to renew its contract with the Maryland State Department of Education for the second fiscal year of the project. A favorable on-site visit and evaluation by the Maryland State Department of Education assured the project of continued federal funding. However, funds could not be granted if the Prince George's County Board of Education did not renew its contract (representing the Growing Together project) with the state board of education.

On April 16, 1974, the Prince George's County Board of Education voted not to approve for fiscal year 1975

ESEA Title III Application "Growing Together." In voting to refuse to continue to accept thousands of dollars in federal funds for such programs as Growing Together, the local county board's conservative majority bloc said they opposed programs that are "innovative" or ones that they believe interfere with "traditional education." The board's action marked the first time a local school board voted to drop a Title III Program before the program actually expired (after its three-year federal funding).

The conservative majority of the board of education had been recently voted into office by a strong, active, anti-busing group called Citizens for Community Schools (CCS). This bloc of conservatives seemed to be intent upon the dismantling of any humanistically oriented, federally funded programs which were a threat to their more conservative-traditional view of education in the public schools. A more thorough analysis of the political ramifications of this decision will be discussed in the last section of this paper.

Several attempts were made by concerned educators and parents with the support of local politicians and interested citizens to reinstate the project. Members of a group of over 80 parents, along with professional educators and concerned citizens, pleaded publicly with the county board of education at an open citizens' participation forum—but without success.

Due to the overwhelming support and pressure in favor of the project, one of the board's liberal members (who voted in favor of the project's continuance) managed to bring the project up again on the board's agenda some three weeks later. But it was again voted against by the board's conservative majority bloc.

At the end of its first year of operation, Project Growing Together was forced to shut down. The project's equipment legally belonged to the county. It was subsequently marked ESEA Title III (for the small possibility that further action would reverse the county's decision) and was distributed to various parts of the county. Most of our equipment was specifically geared to children ages two to five, to whom the county board explicitly declared it was not responsible. In all probability this carefully purchased equipment—purchased with taxpayers' money—is in storage.

The parent advisory committee of the project and State Senator William Goodman petitioned the Maryland State Board of Education to use its authority in reinstating the project. They felt, along with many other parents and educators, that the local board acted irresponsibly and should be forced to fulfill its three-year contract of the project.

The Maryland State Board of Education has responded by designating a hearing. At this time the local board of education must justify its decision to discontinue the project. Several educators, university professors, and parents will also be present to present their support of the project. Reinstatement of Project Growing Together will then be considered by the Maryland State Board of Education. This would be an unprecedented action by the State Board of Education.

DESIGN OF THE PROJECT

Project Growing Together was awarded one of the top priorities in the State of Maryland for its innovative aspects of early childhood education. Being able to identify the elements of a good early childhood program and actually being able to carry out these elements are two separate problems. The more innovative a project, the more time and planning is required to design specifically how these innovations will be carried out.

The staff of the centers have played an important role in determining how to plan and implement the criteria of sound early childhood development. The development of a team-teaching approach has been conducive to this input from the teaching staff. This project was an ongoing learning process for all of us involved in the project's implementation.

While a great deal of expertise has gone into detailing the elements of a good early childhood development program, it should be mentioned that a great chasm exists between being able to identify the elements and actually implementing them.

The next section will include an outline of project objectives and specific innovative aspects which proved to be significantly beneficial.

Table 4–1
Project Growing Together: Objectives

1. To establish two early childhood development centers
   a. To provide experiences and activities for young children (ages two to five) which will contribute to their development in the areas of:
      1. language
      2. socialization skills
      3. perceptual and motor skills
      4. self-concept and emotional adjustment
      5. cognitive formation
      6. problem solving and decision making
   b. To help in the identification of problems at an early age by observations and screenings, and to follow up on such identification with appropriate referrals and contacts with agency consultants.
   c. To expose children at an early age to others from a variety of socioeconomic levels and from different ethnic groups.
   d. To provide opportunities for including mildy handicapped children in a day-care center with normal (nonhandicapped) children.
2. To involve parents in the programs and activities which address their children's needs
   a. To establish parent advisory groups for each center.
   b. To involve fathers and mothers in parent discussion groups and parent education programs.

Table 4–1 (Continued)

    c. To use parents as volunteers, practicing their skills in working with children and with a team of professionals.

    d. To provide opportunities for parent-child relationships through parent involvement with the center.

3. To involve the community in the operation and appreciation of the rationale for early childhood development programs.

4. To meet the need for training in early childhood programs by

    a. The development of a team approach to teaching young children in open-space areas.

    b. Cooperative arrangements for in-service training and credit for career development.

    c. Providing experience for student teachers in a comprehensive early childhood development center.

5. To develop a model of interagency cooperation

    a. To establish advisory groups with representation from health, education, social services, the University, and parents.

    b. To share resources and personnel in an innovative way.

    c. To eliminate costly duplication of staff and efforts.

Continuous assessment of progress toward the program's objectives will be achieved through:

1. Project organization and management that provides for

    a. Clear statements of objectives that are known to all staff and consumers of services.

    b. Definitions of roles and assignment of responsibilities of staff and parent and community participants.

    c. Definitions of lines of authority among staff and participants.

    d. Attention paid to relationships among staff members, development of a team approach, commitment to open communication, concern for feelings, supervisors who listen, staff members who speak out.

    e. Regularly scheduled group staff meetings on a basis of open discussion, mutual respect, decisions arrived at by consensus rather than by decree.

    f. Individual conferences of supervisor and staff members around professional assessment and personal concerns as appropriate.

2. Monitoring of program development through:

    a. Continuing anecdotal records of individual children's progress on a time-sampling basis.

    b. Observations by members of evaluation team.

    c. Parent-teacher conferences.

    d. Records of parent involvement.

    e. Opportunities for feedback from parents on both formal and informal basis.

SUBSTANTIALLY BENEFICIAL ASPECTS

Several aspects of the project proved to be substantially beneficial. These aspects are discussed below.

Team Teaching and Open-Space Areas

Our first experimental approach to this idea was developed during preservice workshops. A group of teachers went to observe open-space classrooms in other schools in Prince George's County, with the idea in mind of developing a model of their own. Our model was based upon these observations, some reading in the area, and a great deal of brainstorming by the team. Input from all individuals involved was essential to the development of a team which democratically made decisions and participated equally in implementing their curriculum.

At the Gaywood Elementary Center two classrooms began working together in this team approach. The equipment in one room was designed for large-muscle development and more "expressive" kinds of activities (such as a music and movement area, a housekeeping area, a puppet theater area, and areas for a larger climber and slide, blocks, and a workbench).

The second room was designed for small-muscle development and quiet kinds of activities which concentrated more upon cognitive growth. Interest areas in the "quiet room" included a book and music area, a large science table area (with pets, insects, and plants), an art table (where cooking was also done), and an area for table games.

Each room was staffed with a full-time teacher and teacher aide, with another teacher aide going from classroom to classroom where needed. Instead of 15 children in each self-contained classroom, a group of 30 children shared two rooms. Children were free to go to the room and area of their choice. If teachers felt there was the need for further individual development in certain areas, certain children were encouraged to explore these areas.

Because the organization of activities is essential to smooth functioning, teachers used the daily rest time for planning. Large-group activities were planned by different teachers at different times. Each teacher brings individual skills and resources to the team; in this kind of approach, all the children can benefit from their combined pool of resources.

As the open-space idea proved to be so beneficial, other classrooms in the project began to co-op resources, teachers, equipment, and classroom space. This approach was an extremely successful and innovative way of structuring a classroom to provide individual independent activities for children and to meet the individual needs of a large group of children. Areas of interest were organized in such a way that children learned to regulate themselves and to take much more responsibility for choosing what they themselves wanted to do. Far more physical development was made possible due to the space for large climbers, workbenches, and a large area for block-building and water play which were always available for use.

A great deal of space and money was saved as equipment did not have to be duplicated and "fit into" each room. We could therefore provide for more kinds of activities and have enough space in which to thoroughly enjoy them.

Parent Involvement

The involvement of parents in center activities which address their children's needs is essential. A major goal of this project was to help parents see themselves as their child's first and most important teachers; to support parents in this role; and to help parents broaden their skills in guiding their children's behavior. Many of the ways in which we attempted to achieve this goal were outlined in the project's objectives and carried out as intended. However, one particular aspect of parent involvement is worth mentioning here in more detail.

When we were having financial problems and were unable to hire the people we needed to properly staff our centers, parents enthusiastically and faithfully volunteered on a regular basis. The fact that they felt vitally needed at that particular time was an important element in their involvement. This factor should be taken into consideration in future projects concerned with eliciting parent involvement.

A well-planned and organized schedule for *when* parents are needed and *what in particular they can offer* to do is very important in getting their participation. In this process, we as teachers got to know the parents and therefore were able to better understand their children. The parents, in turn, learned to appreciate the center and the group of people living with their children more.

Pupil Diversity

In both rural and suburban environments, good day care is needed for all children. An important aspect of our centers was their availability to children from varying socioeconomic backgrounds. An analysis of our enrollment by family income level, race, and sex showed that our families came from a variety of socioeconomic levels and different cultural backgrounds. However, upon observing children at play in the centers these differences were not so apparent.

Unfortunately a great disservice is done to children with the economic stigma attached to some day-care programs. Young children are frequently separated along economic lines, those from well-to-do families attending nursery programs with others from similar backgrounds, those from welfare-level families in day-care centers which do not include significant numbers of children from varying backgrounds. Low- and middle-income families find it very difficult or impossible to afford day-care services.

There are important aspects of the process of socialization which are not available to young children in many communities. A child-care center which is intent upon mixing children from varying backgrounds is a fine place to start breaking down prejudices and preconceived attitudes of others who are "different." This is accomplished by allowing children from these varying racial backgrounds and socioeconomic levels to develop as individuals who are capable of understanding, tolerating, and appreciating the traits of all people.

The Handicapped

Several young children with minimal disfunction were referred to the Growing Together program by the offices of special education, pupil services, and the Prince George's County Health Department. We had established policy to make exception to our regular geographic guidelines for enrollment in the cases of children with "special problems." Over 10 percent of our children enrolled were considered to be in this category.

Providing the opportunity for the interaction between handicapped and nonhandicapped children is viewed as especially important for this age group. The importance of mainstreaming children with special problems has been well recognized by early education specialists. It is important for these children to be able to develop by "modeling" other children in normal environments. Furthermore, it is important for children in normal environments to learn to understand and to accept children who may be having trouble learning.

Teacher Selection

Selecting teachers well trained in early childhood education was one of the major factors in being able to provide
for a sound educational opportunity for the children of our
project. Adults with a firm understanding of child development and early childhood education curriculum are more
capable of providing a stimulating and progressive educational environment which emphasizes equally cognitive,
emotional, social, and physical development.

Training Period

A two-week training period for all staff was held prior to
enrollment of children. This gave the staff an opportunity
to explore different approaches to early childhood education and to develop, as a group, a more clear and consistent
direction, philosophy, and curriculum for the project. Input from the University of Maryland, Prince George's
County Child Study Office, and other experts in the field
of education helped make this an extremely good group
learning experience for the staff and helped to establish a
sound framework for curriculum and further staff development.

Adult-Child Ratio

Low adult-child ratio made it possible for individualization
of instruction and more personal guidance for each child.
As the early identification of special problems was a significant component of the project's program, our existing low
adult-child ratio made it possible for us to both identify
individual problems and to respond to the individual needs
of all our children.

Evaluation Process

Formal evaluation of the child development aspect of the program was carried on under the direction of Johns Hopkins University in cooperation with the State Department of Education. Pretesting of the children included subtests from the Peabody Picture Vocabulary Tests, the Stanford Binet Intelligence Quotient, and two self-concept measures.

The project was terminated before we were able to perform the post-testing, which would have shown how the children of the project progressed. Therefore we were unable to prove statistically the educational and developmental value of our project. This was a very frustrating fact to accept for many of us who were convinced of the project's critical worth and value to the total development of the young child.

We have learned through this experience the critical importance of having a statistical analytical evaluation of project objectives in order to prove the value of a project. Had we been given the opportunity to postevaluate our children's progress, we would have had far more convincing proof of the project's value.

PROBLEM AREAS

Participating in innovation and change in education is an exciting and challenging endeavor. But it also requires a great deal of creative insight, patience, and hard work to succeed. We will now look at some of the problems we had in implementing various innovative aspects of our program. Recommendations are made as to how one might alleviate these problems.

Community Support

When it became evident that the project would be terminated, the parent-advisory committees at all three centers and active staff members made several attempts at gaining community and political support; at making people who represent the community aware of the injustice that was being committed in public education.

Letters were sent to local political people who held records of political work in education. Various local organizations concerned with the education and welfare of children were contacted for support. Letters were written to the editors of local newspapers. We sought coverage in all local media—radio, television, and newspapers. Several newspapers responded in support of the project and did feature articles exposing what the county board of education was actually doing and the implications of their action.

These efforts, although not ultimately successful, did help educate and involve the community and accomplished, therefore, one of our major project objectives: to involve the community in the operation and appreciation of the rationale for early childhood development programs. Furthermore the necessity of community involvement and support in the success (therefore continuance) of the project became apparent.

Community support is a critical element in the survival of an educational innovation in the public schools. It is essential that projects dependent upon public funds make concentrated efforts from the onset of their operation to contact local organizations and key political people. Inform them of your program; educate them as to its educational value; elicit their support. Involvement of the community is the key to the successful continuance of such programs.

Parent Involvement

Parent involvement is integral to the previous discussion. The support we received from parents when the project was voted to be discontinued made us aware of the critical importance of parent involvement. Many of us on the staff, however, would like to have had a wider range of parent participation and involvement in the day-to-day workings of the center. The following recommendation is one way of ensuring that all parents become, in some way, involved in their children's center.

Upon registration the parent would sign an "agreement for participation in the school" form—for specifically four hours a month. This participation could be in the classroom or in other activities related to the center. Each teacher would be responsible for keeping a chart of the number of parent hours and also for reminding parents that "their time has come" if they have not lived up to their agreement. In order to ensure that this goal is achieved, parents would pay a fee of $4.00 per month extra which would be refunded at the end of each month only if they participated.

It would be up to each teacher or team of teachers to agree upon what constituted participation. Staff could discuss and share with each other ideas they may have—or that parents may offer as suggestions. Participation in the classroom would certainly be encouraged. Other activities such as repairing school equipment, telephoning other parents about meetings, preparing refreshments for meetings, babysitting for other parents, fundraising, and so on should be included as alternatives.

Personnel

The process of hiring personnel is a good example of the problems we encountered in working within a large bureaucratic school system. Standards and pay scale for public school teachers are different from the standards and pay scale for teachers of most child-care projects. Some of the teachers we hired had worked in the public school system for a number of years, and therefore required, according to system standards of yearly increases, a much higher pay than teachers just beginning their careers.

However, their experience in the public school system was not specifically in early childhood. Therefore we didn't feel they were any more valuable than the newer teachers, nor that they deserved a higher salary. The newer teachers, who were paid less, generally worked harder, and in many ways were more valuable to the project.

More than 75 percent of our budget was allocated to staff salaries. It would save a considerable amount of money if all teachers were paid the same starting salary. When the project continued the second year, they would receive the same increase. A second rate of pay which was scaled lower included teacher-aides, many of whom were working on degrees in school part-time. Their salary should also begin at the same level and be increased the following year at the same rate.

Furthermore, the hiring of staff should not be done all at the beginning of the school year, for enrollment is usually slow and staggered in a new center. Paying expensive personnel when they are not needed can be another unnecessary expense. Staff should be hired only when needed.

The training session is an important element to a coherent understanding of project objectives. Two separate training sessions could be held—one in the beginning of the year and one approximately six weeks later (or when needed for new staff members). Tapes of the initial training sessions could be used. The second training period could be held in the afternoon during "rest time" so that original staff members could also participate.

Staff Facilitator

Poor communications between staff members is often a problem which can fester into creating an intense and unpleasant environment. Money should clearly be put into the budget for an outside, objective, and well-trained staff facilitator to conduct staff group dynamics sessions.

It is strongly recommended that this staff facilitator be hired from the onset of the project. This person could conduct a group dynamics workshop in the two-week training period, thus providing some initial mutual understanding and sensitivity among staff members.

A contract should be worked out in advance clearly defining expectations and responsibilities of the staff facilitator such as dates of meeting, an agreed-upon fee, and any other extra duties. As consistency is important, this contract should extend throughout the school. Good communications are critical in establishing unity and solidarity among staff members and in establishing a good feeling about being an active and participating member of an organization.

Project Implementation

Due to the complexity and variety of project objectives, innovative features should be implemented gradually and according to a preplanned time frame. It should be made clear who in particular (the director, coordinators, teachers, steering committee, or parent advisory group) is responsible for specific aspects of implementation. Project objectives and activities should take place according to a specified time frame.

The necessary planning and preparation prior to the opening of the center for children of Project Growing Together was cut in half, causing project operation and administration to be a staggering and unorganized responsibility. When working within a large school system or other large bureaucracy, the project director and coordinators should have at least two uninterrupted weeks of orientation of the various departments and their functions within the system. Administrative procedures and policies should be clearly explained. Only then can intelligent planning and implementation begin.

During this period of administrative planning a project brochure should be constructed defining enrollment fees and policies, sick-leave policies for children, vacation days, half-day enrollment, and hours of the center. A brief summary of project objectives, philosophy, and any other school procedures should be clearly outlined before enrollment begins.

POLITICAL RAMIFICATIONS

We will now turn to a discussion of the problems of working within a large bureaucracy and the political influences which determined the discontinuance of Project Growing Together.

A bureaucratic administrative structure can either promote or detract from good child-care programs. Working within a large bureaucratic school system was one of the administrative staff's most difficult, frustrating, and pervasive problems, and influenced, on all levels, the way in which we were able or unable to accomplish certain project objectives.

Project Growing Together was initiated and designed by a group of individuals representing different bureaucracies and differing interests. "Interagency cooperation" was, to us, in reality an interagency nightmare! The problems of working in a single large bureaucracy are well known. However, our difficulty was working with several different bureaucracies—one of them being the tenth largest school system in the country.

Demands for strict accountability and compliance with complex local, state, and federal regulations forced our director to work almost solely with the problems of red tape, poor communications between agencies, and the slow bureaucratic procedures that prevent decisions from being made and actions from being carried out.

The power base and decision-making process in this project was never clearly defined. Were we working for the steering committee who represented several different agencies, or were we working for the Prince George's County Board of Education? Miscalculations made by the steering committee concerning predicted enrollment caused us to hire all our teachers when they weren't needed. The committee predicted falsely that enrollment would be full, because there was such a demand for child care. Initial enrollment was slow. Twenty-five percent of our budget was dependent upon incoming fees; and the lack of such fees to pay the expensive professionals we hired created a serious deficit in our budget.

This miscalculation caused countless problems later on in the year due to the loss of funds. When enrollment was full and we needed funds to hire more people, there were no funds available. Staff were forced to change hours and staffing patterns many different times, many of us working overtime with no pay to make up for the needed extra adults. This is one of the many examples of policy being set by people who do not have to bear the brunt of the responsibility of that policy.

The degree of autonomy of the director of a child-care project and the amount of community interest and support seem to be the key to successful organization and operation of a project. The degree of autonomy of a project usually depends upon the interest which the superintendent of schools and the political board of education exhibits toward day-care programs, and, furthermore, their willingness to delegate decision-making authority to the center's director. Our director had no such confidence in her decision-making authority.

This was further complicated in our project by a steering committee representing different bureaucracies who were also part of our decision-making process. Decisions to be made for the project's operation had to go through so many different bureaucratic channels that we often felt we would have been better off on our own.

The advantages of operating within a bureaucracy, however, should not be overlooked. Locating in established educational buildings, having access to rent-free classroom space and maintenance, given the opportunity of utilizing existing school resource materials and personnel, and the existence of an established framework through which to operate the project are all worthwhile aspects to consider.

One of the main reasons for our difficulties within this particularly large school system was that it was unfamiliar with early childhood education standards. Our project was the first large-scale early childhood education program in Prince George's County. Instead of the system making adjustments and exceptions, we were usually forced to comply to their standards, making compromises in what we knew would be best for our project.

With the acceptance of public funds one must also accept the strings attached to those funds. We were convinced that a conscientious effort to educate public school officials as to the value of our program would make them more willing to make exceptions to their standards to meet our specific needs.

It should be pointed out, however, that the specific problems we encountered in the Prince George's County school system were, in part at least, due to its size and to the conservative nature of the school board. These problems could be avoided in a more progressive and decentralized school system. The advantages of the latter kind of school system would make working within it far more palatable.

The problems we encountered with the Prince George's County school system lead us to the political reasons for the discontinuance of Project Growing Together. As was mentioned earlier, the Maryland State Department of Education had conducted a favorable on-site evaluation of the project and was prepared to continue our funding. It had, in fact, been so impressed with the project that it had granted us the proposed $100,000 increase for which we had applied.

Title III projects, it should be remembered, are "grants to local educational agencies for innovative and exemplary programs to demonstrate ways of making a substantial contribution to the solution of critical educational needs." Federal funds are granted for a three-year period. After that time period, if the project has proved to meet critical educational needs, it is then up to the public to take over the continued funding. This usually means that the local board of education will be responsible for taking over a project.

In March, 1971, the Maryland State Board of Education declared early childhood education to be of high priority in the state of Maryland. In a bulletin of guidelines which the state department of education distributed to local school systems, it indicated the critical importance of early childhood education in the total development of the child. The guidelines stressed the value and saving of money involved in investing in preventive preschool education instead of expensive remedial reading programs which are after the fact.

But these guidelines seemed to fall on deaf ears in Prince George's County. Early childhood educators will be familiar with this frustrating problem: How does one go about educating the public as to the critical importance of the early years in a child's social, emotional, physical, and intellectual growth? The threat that we posed to the Prince George's County School system was that we were becoming successful in this endeavor. One pilot project, which was costing them nothing, could potentially cost them thousands of dollars. As the public was beginning to recognize the value and importance of such programs it would eventually rise up and demand this right to quality education for their young children—an expensive demand.

But the county had other ideas of what to do with its money. It couldn't see the value of making the initial investment in a child's learning process. They instead emphasized getting back to the 3 R's and traditional education. Remedial reading programs were being hailed as the answer to the county's educational problems.

The board's resistance to our project was linked to another aspect of its political conservatism. Another issue, which takes on national significance, played an important role in the county board's decision to discontinue Project Growing Together. At this writing, massive demonstrations and violent attacks upon school children are being held by anti-busing factions who have gathered in Boston to protest the enforcement of the federal government's court-ordered desegregation guidelines.

These guidelines, which came out of the Civil Rights Act of 1964, have been largely ignored by local school systems throughout the country, so that now, some ten years later, the federal government has begun to enforce them. In order to do this the federal government is demanding from local school districts "racial balance statistics" to justify their receipt of federal funds. Federal funds are cut if "racial balance statistics" show continued racial segregation.

The desegregation order in January, 1973, involved the transfer of about 33,000 school children and a substantial increase in busing in Prince George's County. The county's anti-busing group, called Citizens for Community Schools, became actively involved in "regaining control of their schools" in reaction to the federal government's continuing pressure to desegregate. This powerful pressure group voted into office that same year the majority bloc of the Prince George's County School Board.

The majority of the school board, who were voted into office on anti-busing sentiment, began to implement their authority by engaging in a "systematic attempt to rid the school system of anything that had a federal label attached to it." (2) Many of us felt that the board's action was part of a campaign to dismantle federally funded programs aimed at easing racial tensions in the county, and was furthermore a punitive measure designed to get back at the federal courts for ordering the desegregation of the county schools.

By ending any experimental programs that could help children of different races get along better, the school board is creating a self-fulfilling prophecy. They always predicted court ordered busing wouldn't work—and they will see to it that it doesn't. The only problem with this scheme is that the children lose out in the process of all this political jockeying.

A third factor involved in the project's discontinuance was the popular view of day-care. Workers in our project learned quickly to erase this word from our vocabulary, as we realized the stigma attached to it. "Day-care," to many, connotes custodial care—babysitting—and is necessary only for the poor. For many families this is the only kind of child care which is available. Project Growing Together was designed to provide the public with a viable alternative —"quality child care"—run and staffed by well-trained, warm, and dedicated workers who realize the importance of all areas of a child's development: social, emotional, physical, and intellectual.

But here, again, we are faced with a political controversy over whether women in our society deserve the same rights of free expression of their worth through meaningful work. We were accused by the anti-busing CCS group of attempting to disrupt the family, of attempting to take their children away from them. We felt our project would help strengthen the family.

Our defense of this misconception was to attach strong emphasis upon the educational component of our project's curriculum. In a sense we were forced to compromise our professional understanding of the importance of establishing a positive self-concept in the young child, of the importance of developing a creative, inquiring, open mind—to the priority of building intellectual skills. We had to sell our program on a slate that many of us knew professionally was not completely educationally sound.

CONCLUSIONS

In conclusion, the workers in Project Growing Together, through this experience, have become enlightened in understanding the problems that beset our public institutions of education. We have seen the degree of abused political power that determines our public educational priorities. The problems of precarious funding and of public misunderstanding as to the value of quality child care for the future of our society must not discourage us. We must continue to organize, to educate, and to build.

We must maintain a historical perspective as well. The right to quality education, the right of all children to quality child care and health, the right of women to seek meaningful work outside the home must be fought for—just as all other rights in this country have been fought for. The people must be organized and educated as to the critical value of innovation and change in our public institutions of education. The people must be organized into realizing the importance of demanding the rights of equality for women and children. Our commitment to this ideal has become strengthened. We must remain firm in our work and determination to go forward.

The only way these rights can be achieved is through the passage of a comprehensive child care bill, similar to the bill vetoed by President Nixon in 1971. Funding on a national scale is the only way to ensure quality and comprehensive child care for all.

NOTES

1. Maryland State Department of Education. "Guidelines for Early Childhood Education." *Maryland School Bulletin* 48 (Sept., 1972).
2. Jesse Warr (a local boardmember who was in support of the project). *Washington Post.* May 22, 1974.

Chapter Five

MIRROR FOR ADVOCATES:
THE BERKELEY EXPERIENCE*

Kay Martin and
Mary Millman

Since early 1973, when the bottom dropped out of the day-care boom that began in 1968, we, as advocates for child care, have come more or less full circle. The severe cutback and limitation of Title IVA funds requires that we refocus our energies on the basic problem: the lack of child-care services for those who need them.

A critical look at the Title IVA era reveals two reasons why we are no closer to a comprehensive system of child-care services now than we were in 1968. First, by concentrating on the acquisition of Title IVA funds, we helped to identify child care with a population that was small, politi-

*Materials for this article were gathered between 1971 and 1973. The article was written in July, 1973. While certain changes took place in the Berkeley Child Care Development Council, the Office of Community Child Care, and the general local child-care scene between the date of writing and the date of publication, the authors believe the basic theses still hold.

cally vulnerable, and, most important, defined by criteria largely irrelevant to the needs of children for care. Rather than taking into account the situation and developmental requirements of the individual child—which are the raison d'être of child-care services—eligibility for Title IVA subsidies rests on the economic status of the child's family. We worked with the federal government to reduce the welfare rolls instead of working for our own goal of providing care for children. Secondly, though we organized into collective action groups, we allowed these groups to accept the program of coordination, handed down from the federal government for the ostensible purpose of making the existing system of services more rational and efficient, without remembering that it was the very lack of services that called us together in the first place. We got entirely sidetracked from our main concern.

In general, child-care advocates have failed to keep in mind the basic principle upon which organizing should be based and the standard by which progress should be measured. The needs of children are the only legitimate organizational basis of any collective child-care advocacy group. If this principle were acknowledged, it would be impossible to conceive of child care as restricted to any one group of families or children. All children, irrespective of their ethnic origins or the income level of their families, have a fundamental right to quality care. This implies, in our heterogeneous society, a broad spectrum of services sensitive to the differing needs of individual children and their families. It means, in practice, that the unattended or isolated child of the middle-class family has as legitimate a need for care as the neglected child of the single parent struggling to survive. However, since no public body has ever taken the responsibility for the provision of universal day-care services, collective action groups must organize with this as their goal. The measure of their success will be increased day-care services which offer quality care to all children.

The experience of our community—Berkeley, California—illustrates the consequences of the failure of collective advocacy to remember the needs of all children and the corresponding failure to measure progress by a concrete increase in services. Over the past five years, Berkeley's collective action groups exerted so little influence in the provision of services that the overwhelming majority of new child-care slots issued from the efforts of people who eschewed the process of collective advocacy altogether. They focused instead on the provision of direct services for that particular segment of the population they knew well enough to serve. This failure of collective advocacy need not have occurred, nor does it argue against collective action per se. Based on three years involvement with the Berkeley Community Coordinated Child Care (4-C's) organization, we have come to believe in the need for collective organizations devoted to child-care advocacy, so long as the purpose of such organizations is to create programs where there are none.

Collective action is the only avenue to comprehensive services, since spontaneous growth, however vigorous, will necessarily result in piecemeal development: providers have to limit the scope of their vision to act at all. For people who cannot afford to purchase what the market provides, and for people whose needs have been systematically ignored by subsidy programs, collective action is the logical means to make their needs felt. Collective action is also necessarily political. Funding for universal child care is so far from reality in the United States that only well-organized, militant pressure by a cohesive group of potential users, present users, and providers of child care can generate the political energy necessary to ensure subsidization of child-care programs for all who need them.

Berkeley's child-care history parallels that of other communities in that, until the late 1960s, most services had developed as a result of spontaneous individual actions. By the late sixties, however, it was generally recognized that such development had not produced enough services for the city's children. The two citizen organizations which tried to "do something about the day-care problem" lost sight of their only conceivable goal—to produce more programs. The Committee on Child Care (1968–1971) became enamored with rational planning which was supposed to supply it with a program for action and then devoted its energies to implementing the federal government's model for 4-C's. The successor organization, Berkeley Child Care Development Council (BCCDC), also failed to produce new, quality child-care programs, since, after it floundered as a coordinating agency, it got lost in the struggle to create a new city bureaucracy and the search for funds for such a structure to administer. Neither organization succeeded in altering Berkeley's basic configuration of services, nor had any significant effect on the growing number of slots for children in Berkeley. Of the 644 new spaces for child care created between 1970 and 1973, only 55 of these resulted from the coordinating organizations. The administrative costs, which began at $25,000 in 1971, swelled to $100,000 per year in 1972–73. In this latter year the actual money spent on care for children by this expensive administrative agency was only $12,000, though it potentially had $600,-000 at its disposal.

The things that got in the way of providing services for children are the subject of the following history. They are presented in detail because we suspect many individuals in other communities find themselves now, at the end of a long period of activity, perplexed as to why they haven't really made much progress. We think Berkeley's story goes more than half the distance toward providing the answers.

The Committee on Child Care, 1968–1971

In 1968 a group of radical women in league with some professionals from the fields of health, social welfare, and education appeared before the Berkeley City Council and the Berkeley Unified School District (BUSD) to request that an official committee on child care be organized. Previous organizing efforts aimed at increasing child-care services in the city had not been successful, but this group succeeded in elevating the issue of child care to the level of official citywide concern. Recognition was readily granted by the city and the school district, but they blunted the original radical makeup of the committee by appointing liberals and professionals. The committee was charged "to study the need, make a report on the need, and line up funding resources to meet the need." The actions of the relevant bureaucracies belied their apparent supportive position: they offered to their newly recognized Committee on Child Care neither money nor staff. Office space and supplies were donated by the Early Childhood Education Department of BUSD.

Official recognition was a critical step for the child-care forces in Berkeley because it determined the nature and parameters of action for the next four years; the committee prematurely committed itself to work within a system which it ill understood. The committee did not clearly envision that it might have to become an independent alternative to the existing bureaucracies in order to increase services. The city manifestly did not want the responsibility for child care and embraced the committee as a way to deflect demands for services. The school district, which by historical accident was the provider of the most extensive child-care services in the city at the time,* was the true locus of power

*California was the only state in the nation to maintain the Lanham Act child-care centers of World War II. These were operated and administered by the local school district. In Berkeley there were 125 children enrolled in 1965 with a waiting list of approximately 200 children.

in the city's loose system of services. The Committee on Child Care never located itself properly with respect to these two bureaucracies; it did not anticipate that to increase services it might have to bludgeon the city into appropriating funds, and it did not realize that to accomplish comprehensive planning it would have to become the antagonist of BUSD's vested interest in its own service programs and the bureaucracy that ran them. Without funds or staff, official sanction rendered the Committee on Child Care a beggar to its appointing organizations. Though the committee acted out of genuine altruistic concern about the needs of children and thereby steered clear of the rock of excessive concern for its own power and status, its civic-minded, liberal disposition thrust it into the whirlpool of rational planning to solve the problem that it had volunteered to "do something about."

In its first year of meetings, the Committee on Child Care could not settle the central question of the exact task to which it should devote its efforts, and consequently there was no settlement on strategy. Committee discussion posed the question of how to combine effective program development with long-range planning in either/or terms. One faction, wary of highly paid consultant firms which produced words but not programs, urged the group to start services immediately. The other faction wanted more information. They knew rough statistics on need from three previous local reports and could deduce from Department of Labor statistics that 7,200 of Berkeley's working women had minor children who needed some form of care, but the group was not sure that it really knew the scope of the problem. Rational planning, which they believed would take into account the "diversity of needs and resources, as well as problems of financing and administration," would lay bare the anatomy of the city's child-care needs. An incidental benefit would be the reconciliation of the committee with its calling, for as the faction that wanted a plan became dominant and the activists withdrew, the committee came to believe that a plan would provide a blueprint for action that would produce services. Pressures from the city also came into play, for the city manager let it be known that the city council certainly would not act without a plan, though no guarantee was given that the plan would compel action. So the Committee on Child Care, after fishing about unsuccessfully for foundation funds, returned to the city council and the school district to request a joint appropriation for a comprehensive plan for child care in Berkeley. The contract for such a study was finally awarded to a local research and development firm that specialized in technical information for poverty agencies in February, 1970, more than a year and a half after the committee itself had been charged to "study the need."

Before the year was out the committee and the community received their plan. *Care for Our Children: A Comprehensive Plan for Child Care Services* purported to pinpoint the "need" for day care and to outline in a series of specific proposals a "blueprint for action." Calculating from the results of a poorly worded survey questionnaire, the plan identified more than 2,000 children who had no services of any kind available to them and projected a low estimate of the number of preschool children needing full day care at about 650. The blueprint for action, however, did not directly address these quantitative needs. Proposals for a series of programs designed to fill the gaps in the existing spectrum of services were presented, but the plan did not include a strategy for linking these model program ideas to the actual children needing care. Infant care, preschool and primary care, upper elementary care, care of the sick child, and supportive services for licensed family day-care operators were advocated. The plan also strongly proposed the establishment of a 4-C's coordinating council with an office of community child care as its agency.

In effect, the plan divided responsibility for the proposed programs between the school district and the city. It placed in the hands of BUSD all full day care for children two-and-a-half to twelve years old but failed to map out feasible methods for moving the ponderous school bureaucracy to "reconstitute" its primary schools into the proposed early learning centers which would combine care and education in a single program. Of the remaining programs which would devolve upon the city, the infant-care proposal had the greatest potential for alleviating need, since it envisioned a cluster of four experimental centers capable of meeting about one quarter of the need in the first year. This recommendation was never carried out, however, because the Committee on Child Care threw its decisive support to two proposals less difficult to implement: a homemaker service for sick children which would function under the administration of the health department, and an afterschool program for fourth to sixth graders funded and staffed by BUSD. The committee itself would merely advise these programs, which were both funded by October, 1971. Members of the Committee on Child Care thought that they could "develop" these two new services at the same time they pursued what had come to be their main concern—the establishment of the Office of Community Child Care and the 4-C's Coordinating Council.

The Comprehensive Plan instructed the Committee on Child Care to convene a "representative group of parents and public and private agencies" to form a 4-C's, which, with an office of community child care, would become the "hub of a system of child care services, linking all the separate, autonomous programs into a working whole and providing the mechanism for development of much needed new programs." Incorporated and independent, the 4-C's organization would be totally devoted to collective advocacy for child care. Its staff would translate the power of the council into action. To the Committee on Child Care, weary from doing the multitudinous tasks required by the recommendations in the plan that they had agreed upon and unwilling to volunteer their services much longer, the creation of the Office of Community Child Care seemed a higher priority than the creation of the policy council. Accordingly a day-care coordinator whose job would be to establish a day-care referral service was employed by BUSD toward the end of 1970. By the spring of 1971, a program developer whose specific charge was to implement the proposals in the plan was also hired. The fledgling Office of Community Child Care began its work in a corner of a room in the health department, with little supportive publicity to announce its creation. It was the staff of a council which didn't exist. The idea of a community child-care agency taking its policy direction from a representative council was betrayed by the sequence of events. The Office of Community Child Care was hired and paid by the school district, to which it therefore was responsible.

As the Committee on Child Care approached the process of interviewing and deciding on the staff of the office, it was brought face to face with the problem of its own composition. Members became acutely conscious that the dominant image of their committee was white middle class. Although the committee and the comprehensive plan had consistently paid lip service to the goal of child care for all who needed it, in practice the focus was narrowing to services for low-income people. Committee members believed that the majority of low-income families were black and reasoned that it was socially important to hire black people to fill the two available positions.* But in their concern for social justice they twisted the logic of the civil rights move-

*Berkeley (population: 120,000) was the first northern city to adopt an integrated school policy achieved through citywide busing (1963). The political tenor is generally progressive, and by 1971 there were several vocal and politically articulate poverty agencies. The geographic limits of the Model Cities area demarked a relatively large part of the town.

ment (in which many members had been active participants) by neglecting to enforce basic standards of
competence in their choice: they selected for program developer a black male who had far less experience in the area
of child care than several highly qualified white female applicants. This decision installed a syllogism in Berkeley's
child-care structure. Day care is for the poor, the poor are
black, therefore day care is for blacks—the eventual consequence would be the preoccupation of the policy council
and the office with welfare-related day care for black families to the exclusion of comprehensive child care for lower-
middle and middle-class families of whatever race.

The committee's sensitivity to its community image
also bore heavily on its role in the creation of the 4-C's
council, for the committee decided it was not representative enough even to perform the facilitating function of
convenor. Accordingly it called together day-care users and
agency providers into a child-care steering committee
which proceeded to form a council more representative
than the guidelines of the federal 4-C's program suggested.
The steering committee settled on a 29-member council of
which parents would comprise the majority (not the suggested one-third) of the membership. Four of the parent
seats were reserved for people who needed child care but
could not find a program to fit their needs. These "potential users" would embody the conscience of the council;
they would be a constant reminder of the city's large unmet
need for day-care services.

The first meeting of the Berkeley Child Care Development Council (BCCDC) in June, 1971, marked the birth of
an organization with congenital defects. The steering committee, in its eagerness to create a perfectly representative
council, allowed several critical issues to fall by the wayside.
Little attention had been devoted to defining the tasks of
the council and the staff: were they to develop new services,
coordinate what already existed, or both? If the latter, how

were these activities to be meshed? No one attempted to formulate the relationship between the council and its staff. It was nowhere precisely spelled out to whom the staff would be accountable, or indeed, for what. In steering committee meetings, when the newly hired program developer was unable to provide information which members had asked for, no one was willing to confront the larger issue of accountability. The carryover of members from the steering committee onto the BCCDC ensured the perpetuation of these ambiguities.

Once the BCCDC had been established, the Committee on Child Care closed up shop, believing that the new coordinating council would guarantee a rational future for child-care services in Berkeley. Everyone assumed, on the basis of federal literature about 4-C's and the advice of the comprehensive plan, that when all those concerned with child care got together to talk about their interests and problems, they would act together to effect solutions. But the crucial link between discussion and action, absent from the literature, was no more in evidence when the actual council met. BCCDC succeeded in collecting most of the key people: the Alameda County day-care coordinator (conduit for Title IVA funds); the program director of the University of California child-care center serving student parents (pioneer of campus day care in the state); the school district's acting director of early childhood education; the Model Cities board member whose main interest was child care; and a high ranking official in the city's Social Planning Department, not to mention the majority of parents and users of child-care services. While none of the agency people were empowered ultimately to act for their bureaucracies, they were all knowledgeable and influential. Yet collecting them to talk not only failed to produce genuine program development, it failed to accomplish even minimal coordination among existing programs. The school district representative came to council meetings and announced BUSD's plans for expansion of its own programs rather than submitting the content of the plans for the collective scrutiny and advice of BCCDC. The campus child-care representative ceased to participate because she could provide more child care by devoting the time to her own program than by attending BCCDC meetings. The Model Cities representative, also chairman of the council, worked to prevent Model Cities and BCCDC from collaborating on the development of programs in the Model Cities area. The main concern of the city liaison was to upgrade the status of the staff of the office and to deflect whatever criticism might be directed at their performance.

BERKELEY CHILD CARE DEVELOPMENT COUNCIL (1971–)

The establishment of the Berkeley Child Care Development Council (BCCDC) marked the second phase of that process of collective action which found its ultimate justification in the increase of direct child-care services in Berkeley. But like its predecessor, the Committee on Child Care, the BCCDC failed to accomplish the task because it never used the provision of services as the measure of success. Whereas the Committee on Child Care had retreated from the responsibility for producing programs into the more "professional" role of a planning organization, the BCCDC would forsake its charge in the pursuit of power, control, and money. The BCCDC and its agency, the Office of Community Child Care, became adroit at exploiting the garbled ideology of 4-C's to obscure its failures. Though neither coordination nor program development were being carried out, the organization claimed to do both; when challenged about one, BCCDC claimed to be busy doing the other.

In the first six months of its existence the BCCDC established a reputation for doing nothing. The agency representatives, with their strong and often conflicting interests, dominated the naïve parent majority; and the only common task was drafting bylaws and articles of incorporation so that the organization could receive and disburse funds for children's programs. This activity provided the first test of the undefined relationship between the staff and the council. The staff (the program director and the day-care coordinator) hired by BUSD, which essentially left them unsupervised, was not anxious to have its autonomy restricted by accountability to BCCDC. BCCDC was itself so divided and unclear as to its goals that it could not exercise effective authority over its staff. The staff did not step into the vacuum to exercise leadership, and in fact even failed to perform the routine daily work which could have welded the policy council into a coherent organization for the advocacy of child care services.

By spending half a year to write its bylaws and getting bogged down in recriminations about who should do the work, the BCCDC and its office made a tacit decision about priorities; direct services once again came second. The organization inherited responsibility for two direct service programs from the Committee on Child Care—the home-maker service for sick children and the afterschool program. Yet it neglected to monitor these programs, with the result that the health department's desultory administration of the homemaker service went uncorrected and the afterschool program nearly closed because it had no support in its battle with BUSD and the city recreation department for a site. Although no money had been allocated to BCCDC by the city or BUSD for program expansion during fiscal year 1971–72, the organization did not pursue the program development necessary to prepare for the July, 1972, city budget hearings. There was no publicity, no investigation of the city's needs, and no exploration of existing or potential resources.

Mounting pressures for results descended on the organization from three directions in the spring of 1972. The city strongly suggested to BCCDC that, if it intended to be included in the municipal budget for 1972–73, it develop a master plan to justify requests for individual program allocations. At the same time, newly elected potential users on BCCDC who were strong proponents of the view that the only justification for the organization was the provision of new child-care facilities began pressing the staff to proceed with the organizing necessary to generate proposals for child-care programs and to offer the technical assistance necessary to transform proposals into viable programs ready for funding. They were further concerned that the master plan include some criteria for quality upon which BCCDC would base the selection of programs for funding. Just as work began on the drafting of the master plan, a third factor appeared that would rend the organization along racial lines and alter its character altogether. The County Welfare Department announced in May that it would now—after six month's delay—act on BCCDC's application for Title IVA funding for the administrative costs of the organization. (The Office of Community Child Care now comprised five full-time employees, because three new people had been added in February with Emergency Employment Act [EEA] funds from the city.) This development meant BCCDC would not have to go as a pauper to the city, since it could command from the federal government three times what local government would invest. The obvious question arose: what kind of office should be funded with this federal bonanza?

The debate about the office quickly became the arena for all the unresolved conflicts in the organization: staff competency, council-staff responsibilities, status of the organization in the city, its role in relation to direct services, the population to be served if programs were developed, which political philosophy in the heterogeneous membership would dominate. At first the discussion took the form of a rational debate about the structure of the office. The central question was: should the office be hierarchically organized with bureaucratic lines of accountability or should the office revert to a laterally organized, non-hierarchical structure? One faction believed that the hierarchical structure would transform the basic concept of the office from a community organizing, child-care advocacy office to an official city agency. As the debate intensified, the conflict of interest between blacks and whites became clear. The blacks united around the concept of an official agency that would provide high-paying jobs for the black community, while most whites argued against this poverty program model and fought instead for a laterally organized office, which they believed would spend less money on administrative costs and more money on direct services.

The vehemence and bitterness of the controversy between blacks and whites supplanted rational deliberation about the issue. Under the leadership of a new black chairwoman, the blacks on BCCDC coalesced into a coherent political force. The vote on the structure of the office pitted faction against faction, and victory went to the hierarchical agency advocated by the blacks. At this juncture, BCCDC's new leadership could not allow the entire organization to participate in writing the master plan, since this would invite a second battle over control. The master plan was finally conceived and written by a small minority of members and nonmembers brought in especially for the occasion. The final document was drawn up to justify the expenditures of the child-care agency which blacks hoped BCCDC would become. By this time, about half-a-dozen child-care programs that could be funded had been located by community meetings organized by potential users on BCCDC. The faction that had advocated the lateral office structure had simultaneously urged the inclusion of these programs in the request for funds from the city. All mention of these specific programs was excluded from the final master plan, and model budgets were put in their place on the theory that BCCDC should not have its hands tied in program development in the coming year. Those who wanted BCCDC to be a city agency with a highly paid bureaucratic staff wanted an uncommitted budget for that structure to administer.

The battle in BCCDC was carried to the Berkeley City Council where the whites succeeded, through sympathetic city council members, in forestalling approval of the BCCDC master plan. They wanted time for the issues of accountability, clarification of status, standards for programs, and purpose of the organization to be resolved jointly by BCCDC and its two funding agencies, the city and BUSD. The black faction viewed this move as a "condescending setback designed to castrate and destroy BCCDC." At the next meeting of the city council twelve days later, the black faction jammed city council chambers with partisans. All attempts by city council members to query the organization on substantive issues were met with cries of "racism" from the audience. An impromptu address by Chairman Bobby Seale of the Black Panther Party was required at one point to cool out the crowd and avert violence in the hall. The chairwoman of BCCDC once again presented the master plan to the city council, declaring herself at the end of the presentation "for black control . . . this is what we're all about whether we want to admit it or not." She explained later to a local newspaper: "The presence of federal funds brought about another attempt to pimp programs designed for black people. If they're going to pimp it, then I'm going to be in control and decide how many pimps we're going to have in each program."

The black leadership of BCCDC claimed it had bludgeoned the city council into funding child care for black children; in fact, the pyrotechnics in city council chambers accomplished no more than ratification of black control of BCCDC which was, by this time, redundant. It took six months of maneuvering to convert the city council's approval of the master plan into a monetary appropriation of $168,000. The city reserved the right to approve every program allocation separately, and further insisted that municipal funds be released only when matched by Title IV-A. The BCCDC therefore became the pipeline between the programs it selected for funding and the county welfare bureaucracy which controlled the flow of Title IV-A monies. Since BCCDC continued to ignore technical assistance to programs, some of its chosen services did not meet federal quality guidelines and therefore could not accept matched money. Other programs were merely paper proposals with neither staff, site, nor children. BCCDC formed a close alliance with the county, ostensibly to facilitate the release of federal subsidy. But Berkeley's child-care organization had little power in this larger theater and quickly succumbed to the county's stream of excuses for not releasing funds. At the end of fiscal year 1972–73, BCCDC had managed to match about $3,000 of the $168,000 local dollars, "purchasing" therewith 55 child-care slots—less than half of these offering full-day care—from local providers of services. BCCDC came to act essentially as the local screening agency for the county welfare department by identifying places in Berkeley programs for Title IV-A eligible children and by referring this selection to the county for subsidy. The organization that began as a coordinating council and fought to become a city agency had finally abandoned its responsibility to meet the need for child-care services in the city of Berkeley.

POSTSCRIPT: COMMUNITY-COORDINATED CHILD CARE

Although child care became an area of concentrated national attention in 1968, the federal government had no intention of launching universal, public day-care programs. Government policymakers saw day care only as a means to break the poverty cycle and reduce the welfare rolls. Their limited concept of the purposes of day care complemented the concern of many individuals that the federal government not be given license to intervene in the lives of young children. In practice, child care became defined primarily as the problem of the local community, which by the late 1960s, had assimilated the rhetoric of the War on Poverty. Its common catchwords, "citizen participation" and "maximum feasible participation of the poor," emphasized the ideal of local responsibility and self-determination. However, the "community" of the poverty program did not necessarily constitute an exclusive community for rational child-care planning. Community might mean the poor community, the black community, the neighborhood, the city, the county. Yet the need for child-care services cut across these categorical boundaries.

Day-care advocates could not fail to be influenced by the prevailing winds in social-service thinking of the late sixties. One idea, that federal subsidy programs should be restructured to incorporate previously excluded minorities and extirpate institutional racism, was reinforced by the growing third world militancy of the urban ghettos. A second notion, somewhat less conspicuous, saw coordination as the panacea for the ills besetting the social services. Social planners and federal bureaucrats advocated the rationalization of existing programs to meet the needs voiced by the communities. Coordination in its simplest form meant bringing the key actors together on the same game board to sit down and talk to each other. The child-care version of this social policy concept is embodied in the (Community Coordinated Child Care) 4-C's program. The 4-C's was sold to day-care advocates as the solution to the rapidly evolving day-care crisis. The tantalizing vision of a rational system of child-care services advanced in the 4-C's literature seduced liberal, rational, technocratic, and democratic child-care advocates into untold hours of volunteer labor directed toward establishing 4-C's councils in their communities.

Aside from the fundamental irony that communities needed federal dollars for child care and got instead federal guidelines for 4-C's, the concept of coordination itself had serious deficiencies from the outset. The basic question of whether coordination would result in the rational proliferation of services or whether it would act to improve existing services was neither clearly raised nor resolved. Berkeley's child-care organization was virtually paralyzed by this confusion; it was faced with considerable—and articulated—need for increased services, but its informing concept, coordination, sent no clear signals as to how the organization should act in response to this circumstance. The additional failure of the 4-C's concept to deal with the question of whether it is possible to bring all the elements in the existing spectrum of services together in a coordinating structure compounded the confusion. There was no real incentive for the proprietary operator, who received no federal subsidy and who did not need assistance in recruiting children, to associate with a community coordinating effort. On the other hand, for operators of nonprofit programs and programs serving primarily the poor, the expectation of federal funding was a magnet in the center of the coordinating effort. Although 4-C's rhetoric was inclusive —*all* child-care interests should be represented—in practice the absence of a valid organizing goal meant that major groups would be left out. In Berkeley, it was particularly unfortunate that the day-care innovators were never integrated into the coordinating council. Some expressed interest, but they found they could be more productive by concentrating on their own programs.

The federal model of 4-C's would probably be most successful in socially and economically homogeneous communities. But the lesson of the sixties is that most American communities are heterogeneous and many are seriously polarized. The 4-C's failed to offer substantive guidance to these communities where even the task of constructing an organization would be impaired by the multiplicity of interests. And the central problem, how to keep conflicting community interests subordinate to the needs of children, was passed over altogether. In Berkeley three factions became involved in the 4-C's effort: liberal, economically secure volunteers committed to rational procedure within the system; upwardly mobile blacks whose expressed interest was well-paying jobs and status; and parents of preschool children who had an immediate interest in the provision of services. It was assumed that this diversity of interests would orchestrate itself to ameliorate Berkeley's child-care needs. Instead, factional controversy and conflict of interest consumed people's energies. The interest which prevailed reshaped the organization into an agency focused on the needs of a very small group of families, thereby attenuating the cross-community benefits envisioned in the 4-C's program.

Communities bought the federal government's plan for coordination rather than examining their own needs and developing a model of collective action to meet those needs. In Berkeley, and we believe in many other communities, this meant that day-care advocates attacked the wrong problem at the wrong time. Rather than joining together to address the need for child care by developing an expanded and well-articulated system of services, they organized around the concept of coordinating what already existed and were sidetracked even in achieving this goal. The meager results of this focus compel the reexamination of the direction day-care advocacy has taken and strongly suggest the reorganization of our collective advocacy groups into more functional organizations.

All of our experience indicates that those who are seriously concerned about the lack of child-care facilities in this country and in their communities go out and start programs as a first step. This involves a different concept of day-care advocacy than has been prevalent in the last five years. Instead of building an organization from the top down, starting programs should concentrate on the essence of the problem—the needs of children for care. It would also generate a constituency that has a stake in services. Creating direct services may require extremely low-budget or extralegal programs in the beginning, but if the program is viable and worthy of expansion, large-scale operation will probably lead to the search for public subsidy.

At this point, there is likely to be a clear rationale for collective action. We think that starting programs and forming purposeful coalitions is the most sensible activity for communities in the next few years. But local people must not forget that child care is a national issue. The crucial decisions about the shape of the future of child care will be decided in tacit agreements between bureaucrats and administrators unless there is coherent and forceful activity in the communities. The era we face will be the last-ditch fight for a subsidized child-care system that meets the needs of the localities in all of their diversity and that remains in the control of the people whose lives it affects.

BIBLIOGRAPHY

Bourne, Patricia. *Day Care Nightmare.* Berkeley: Institute of Urban and Regional Development, Working Paper no. 145, 1971.

Martin, Kay. *Planning for Day Care.* M.A. Dissertation, University of California at Berkeley, 1972.

Morgan, Gwen. *An Evolution of the 4-C Concept.* Washington: Day Care and Child Development Council of America, 1972.

Pacific Training and Technical Assistance Corporation. *Care for Our Children: A Comprehensive Plan for Child Care Services.* Berkeley: 1971.

Pierce, William L. "Day Care Services: A 'No-Quality' Future?" *Inequality in Education* 13 (Dec., 1972), pp. 49–57.

Ratliff, Patricia. *Organizing to Coordinate Child Care Services.* The greater Minneapolis Day Care Association Early History. Washington: Day Care and Child Development Council of America, 1973.

Robinson, Jini M. "Berkeley Blacks Gain Control of Child Care after Tough Racial Battle." *California Voice,* Aug. 3, 1972.

Chapter Six

ORGANIZING COUNSELING AND COORDINATION: A MODEL FOR CHILD-CARE SERVICES IN COLORADO

Mary W. Van Vlack, M.S.
Ramon C. Blatt, Ph.D.
Paul T. Barnes, Ph.D.

Through our experience with child care programs we*
have developed a model which has been extremely effective
in meeting a wide variety of family needs. At the heart of
this model is the Child Care Counseling-Coordination

*The Child Care Project at the University of Colorado Medical Center, Denver, Colorado, was funded in 1972 by the Office of Child Development (HEW), Grant Number OCD-CB-248. For more information write: University of Colorado Medical Center Child Care Project, UCMC, Box 2338, 4200 E. Ninth Ave., Denver, Col. 80220.

Office (CCO). The CCO offers a resource data bank on day-care and other child-related programs, a day-care referral service, and counseling and support for families. Additional programs and services can be attached to the CCO when the need for them is demonstrated. These may include intervention in day-care arrangements, family referral to other children's services and programs, support and training for family day-care homes and centers, a survey of available resources and needs, impetus for the development of new programs and improvement in the quality of existing ones, and advocacy. This set of services has come to be known as "information and referral" and should be distinguished from the more familiar direct services to children.

This project provided direct services to children in both center and family home care settings, serving approximately 300 children during the 30 months of their operation. The CCO component, however, was able to help nearly 1,000 families (many with two or more children) meet their individual child-care needs at considerably less cost. Our experience with these programs has led us to make a number of specific recommendations for those who wish to provide information and referral services in the area of child care and child welfare.

DEVELOPING A RESOURCE BANK SHOULD RECEIVE HIGHEST PRIORITY BEFORE YOU OPEN FOR SERVICE

The effectiveness of the referrals will always be limited by the extensiveness of available resources. Early in the life of the project we found ourselves open for business with our own child-care center virtually the only resource about which we knew anything. It takes time and effort to pull scattered information together into a usable form.

Directories or Lists of Day-Care Resources and Health and Welfare Services Already May Have Been Developed; Begin with These

A list of the centers and preschools, showing names, addresses, and phone numbers, came with little difficulty from the state licensing agency. Similar lists of licensed day-care mothers came, with greater difficulty, from county licensing agencies. In some of the metropolitan counties these agencies believed it might be harmful to share these lists with a group such as the CCO, and some never grasped the notion that we wished to see new lists which included the names of new day-care mothers every time they were published. These lists, we learned, did not include the names of day-care mothers who requested their names withheld; they also did not, of course, include the names of day-care mothers operating without licenses.

Unlicensed people create quite a problem for an agency such as ours. Where there is a licensing law, their activity is usually illegal and they are difficult to find; but we know that there are many of these people and that many parents routinely rely on them for day care. We have followed a policy of seeking out these day-care mothers and recommending that they obtain a license, suggesting that the cost and hassle may be outweighed by the benefits. This problem has been further compounded by the fact that many day-care mothers do not have licenses because their homes do not meet the requirements; they may live in the central city where their services are desperately needed but where bringing their homes up to the standards is impossible.

We intended for our resource bank to include information on health and welfare services for children in addition to resources for day care. A directory of children's services in the state and a directory of community services in the city, in addition to a phone information and referral agency for health and welfare services, were already available in our area. We began with these, adding our own experience and that which families and other professionals shared with us over time.

CONCENTRATE YOUR EFFORTS ON GATHERING ACCURATE DESCRIPTIONS OF DAY-CARE FACILITIES RATHER THAN ASSESSING "QUALITY" OF PROGRAMS

We have found that we need two types of information about both day-care centers and homes. The first is logistical or business information and includes location, cost, hours of operation, ages of children accepted, and whether there is currently a vacancy in the setting. The second involves information on program and care-giving styles such as: how structured or unstructured, how stimulating, what type of educational experiences, how well-organized and equipped, how warm and loving an environment, and whether caregivers follow "mothering" or "teaching" models. A recurring area of difficulty in any effort to acquire information on child care is quality evaluation. We know of no quality measurement instrument currently available which would meet the rather rigorous demands for reliability, validity, and ease of application that we would place on it. Even if we had such an instrument, we are not convinced that quality evaluation provides the most meaningful and helpful information to assist parents in arranging child care. We have found, instead, that information that describes what happens in settings and what the caregivers are like allows parents to evaluate a setting within the context of their own values and lifestyles.

We may record, for example, that one day-care mother believes children should be toilet trained at 12 months or that another does not spank toddlers but prefers to distract and reorient them. We might describe a day-care center as believing that children should be free to plan their own activities and develop their individual creative talents, and, therefore, every afternoon from rest time until the end of the day, the children have free access to art craft materials without staff interference. We then share such information with parents and offer them some assistance in evaluating it in terms of their own values and needs.

SITE VISITS AND PHONE CONTACTS PROVIDE THE BEST INFORMATION ON SETTINGS

We have tried several techniques for obtaining information. The first effort, in the early stage of the project, was to send a questionnaire to every licensed center and day-care home in the area. Since the original response rate to this questionnaire was only about 64 percent and since new centers and day-care mothers have not routinely filled out the questionnaire, we do not have information from this source for a majority of settings. Even though 64 percent may appear to be a good return for a survey, in the referral business you need access to all the resources. A second shortcoming of our survey developed out of our cooperation with the local licensing agency and another child-care referral agency. To satisfy these people the questionnaire was compromised, discarding items related to program "quality" and leaving only items covering logistical information and items reflecting compliance with licensing regulations. Telephone contacts have provided much better information than the questionnaire. Over the course of the project, CCO staff members have contacted most of the center directors and day-care mothers in the immediate area, some of them almost weekly. These contacts have provided a wealth of the logistical information and quite a lot of information about programs, child-rearing philosophies and attitudes, and problems in the operation of day-care settings. We also organized a program of site visiting to learn more about both centers and day-care homes. We have found this to be the most effective way to learn about setting programs and caregivers' styles, but, because of its sensitive and time-consuming character, this has proven the most difficult to accomplish. Despite several periods of intensive effort we have only visited 30 per cent of the 287 settings closest to the medical center, missing some because of limited staff time and others because some center directors and day-care mothers preferred that we did not visit. Through site visits and regular phone contacts, the CCO staff has built up mutually trusting and supportive relationships with caregivers. These have been invaluable in speeding up the placement assistance process and in opening up and maintaining a network of communication.

In addition to information from surveys and staff contact with settings we have received some consumer-type information from parents. We have recorded the most informative of this. For example, one parent reported to us that one center made her daughter stand in a corner for a misdeed. For some parents this information would rule out a setting entirely, while for others it would sound quite appropriate.

INFORMATION ON RESOURCES SHOULD BE ORGANIZED TO FACILITATE EASY USE AND TO MAKE POSSIBLE REGULAR UPDATING

In a town of 5,000 people the distance between a family's home and child-care facility is important; in a city it is critical. We have, therefore, organized our files geographically. In Denver, zip code areas provided good divisions, but in some areas, larger units, such as quadrants of the city, or smaller units, such as neighborhoods, might be more appropriate. We suggest recording information about each setting on file cards (8" X 5" is a convenient size); we file centers and day-care homes separately by zip code area, each in alphabetical order. Since day-care settings come and go rather quickly, it is essential to have an easy system for adding new cards and removing old ones.

PLAN TO GATHER THE SAME KIND OF INFORMATION ABOUT FAMILIES THAT YOU HAVE COLLECTED ON FACILITIES

Logistical information is critical. Such considerations as age of the child(ren), number of children needing care in the same setting, schedule of hours and days, location, how soon the care is needed, and cost can severely limit the number of settings available to a family. In many situations these may so limit a family's options that it is almost impossible to take other individualistic needs and desires into account. When there is the opportunity—and when a family is willing to—the CCO staff will attempt to learn about such individual factors as: the child's health and development, his unique personality, aspects of programming the family would like to see, how structured a setting, how educational an approach, how important is the homelike environment, and the presence of strains or problems in the family. Then we can attempt to take these considerations into account in searching for settings.

ADAPT YOUR SERVICE APPROACH TO THE NEEDS OF DIFFERENT KINDS OF FAMILIES

We have found that different families respond better to different approaches. For example, one mother who came to us for assistance in finding day care was a very articulate woman who had a great deal to say about her child's development, personality, health, and habits, and who did not feel her needs had been taken into account unless she had the benefit of an extensive personal interview. Another mother who contacted this office was less articulate; she wanted to tell us briefly what she needed and then wait for us to call her back as soon as possible with specific referrals. She would have found the personal interview an imposition and a waste of her time. While this difference implies nothing about either woman's competence or concern as a mother, it certainly demanded different approaches in order to best meet the needs of each woman.

We have handled day-care referral with a variety of approaches: we have made referrals to some families over the phone immediately, without taking time to search for vacancies among resources; we have taken minimal information over the phone immediately, without taking time to search for vacancies among resources; we have taken minimal information over the phone, performed a vacancy search, and then called back with referrals; we have taken extensive information over the phone and done some counseling, performed the search, and called back with referrals; and we have asked some families in our office for an extensive interview, then searched and called them back with referrals. We have found that, when applied uniformly to everyone, one technique is no more effective than any other. It appears that the most successful and most comfortable tactic, both for staff and for families, may be to have available several different procedures for working with different families and to maintain the option of switching to a different approach if one appears to be ineffective.

REFER FAMILIES ONLY TO THOSE SERVICES WHICH CAN MEET SOME OF THEIR NEEDS

Whether referrals involve day care or other services and programs for families, the CCO has evolved a policy of checking out each resource to insure that it offers the particular service the family desires and is actually available to that family. For most requests for assistance in finding day care, a CCO staff member will call resources until he or she can find several acceptable alternatives. In addition to checking for vacancies, this search involves checking to learn whether the age of the child, the schedule of the family, or whatever is acceptable to the caregiver. The staff member may also use this opportunity to share with the caregiver health, developmental, or behavioral information on the child that may have some critical bearing on the success of the day care arrangement. Some families prefer not to wait while the CCO performs this service, but we believe it is worthwhile in most cases and, in a few situations, it is critical.

REFER FAMILIES TO SETTINGS THAT FIT WITH THEIR VALUE SYSTEMS, NOT YOURS

We have found that we are most successful in assisting families when we do not attempt to impose on them our values and preconceived notions of what makes a good day-care program. For each family there appears to be some key or critical consideration unique to them such that, when a setting can be found which meets that consideration, there is a foundation for immediate trust and for a solid working relationship between family and caregiver. This may be a logistical consideration, such as finding someone to care for a child while the parent works at night, or it may have to do with caregiving style, such as finding a grandmother-substitute for a small infant and his uncertain and inexperienced mother. Whatever it is, it will be unique with each family and will be the most impor tant factor for the CCO staff and the family to take into account.

CCO STAFF MUST BE SENSITIVE TO UNEXPRESSED NEEDS AND ABLE TO RESPOND IN A WAY MEANINGFUL TO THE FAMILY AND THE CAREGIVER

Once logistical concerns and individualistic issues are dealt with, families and CCO staff have the option of engaging in more involved counseling and parent education. The staff can help families sort out what they want or need and what kinds of things are appropriate for their child. One family came to us seeking a new child-care center for their three-year-old son. He had not adjusted well in his current center; he cried, resisted going, and was generally unhappy. The counselor discovered that, although intellectually advanced, he was small for his age and was less robust and self-assertive than any other child in his day-care center. The counselor suggested that a day-care home which provided an intellectually enriched milieu, had fewer children, and where there were other, smaller children, might be more appropriate than another center. With a little more growth and with experience with a small group of children, he might make an easier transition to a center or kindergarten at a later time.

The CCO can do more than increase parents' awareness of their options in the community and how each of these may affect the family and the child; it may also assist families in evaluating settings and determining what each will offer. Many parents must be *told* of the importance of visiting any setting before selecting it, of taking their child along for the visit, and, if possible, spending several hours there.

Beyond the counseling and assistance associated with helping families arrange for day care, CCO staff has become involved with families and caregivers in a variety of other ways. In situations where there have been problems in the day-care setting, we have intervened to help everyone develop a better understanding of the problems and have recommended courses of action and possible sources of outside assistance. We have helped parents and caregivers in finding pediatricians, family counseling, and assistance for speech and hearing difficulties.

We have found that, most often, the person who uncovers a problem with a child or a family is the caregiver, who has daily contact and an ongoing relationship with the family. By giving support and assistance to the caregiver, we often have an effective, though indirect, route for assisting the family.

CONSIDER DEVELOPING OTHER COMPONENT PROGRAMS, SERVICES, AND ACTIVITIES, ADDED AS THE NEED FOR THEM IS DEMONSTRATED

We found that the training and support program for family home care mothers, which was originally conceived as working out of our child-care center, functioned much more smoothly when it affiliated with the CCO. Our office assisted in recruiting family day-care mothers to participate in training workshops and worked closely with them through the placement process and, subsequently, when problems have arisen. Involvement with the support program has helped these women develop a multidimensional and more long-term relationship with the project. Through this program the project has also obtained an opportunity to enhance the quality of care provided in these settings. Through the CCO's experience base and through the data we have accumulated in surveying child-care resources and in assisting parents to find resources to meet their needs, we are able to speak effectively regarding what is available in this community, what is lacking, and where the major problems lie. We have found ourselves in the almost unique position of having essential information to share with administrative and policy-making agencies in the area, such as the city council, the local day care licensing agency, our state department of social services, and the state legislature.

The CCO component of the child-care project began as a subordinate unit of our program of direct services to children. It became the project's most important and central component as it expanded to meet the diverse needs of families. We believe that this child-care model is unique in its approach to service, use of resources, flexibility, potential for expansion, and community orientation.

The model is unique in that it offers indirect services to children and direct services to parents and caregivers. It is less expensive than direct services to children and may, in some situations, be a more cost-effective approach for planning the expansion of resources. It is also one of very few efforts to offer parents day-care-related consultative services, supporting and consulting with the adults regarding the care of children and offering direct services only in occasional situations. Since the CCO is in a position to identify problems as they occur, it has the capacity to offer training for primary prevention as well as working with parents and caregivers to avoid the recurrence of problems in the same areas.

Not only does the CCO model effectively use existing resources in the community, but it can also evolve and expand to provide direct services to children, to stimulate others to develop such services, to work on child-care legislation, or to reach into a wide range of additional activities. Its capacity to expand is limited only by the needs it identifies and the funding it obtains.

The CCO model is also unique in its capacity to assist a wide range of families and children. Unlike a direct service model, it is not committed to offering specific types of programs to a specific group of children. It offers access to a variety of settings and, once a type is selected, some choice of settings of that type. The CCO's services are not limited by children's ages, family income, cultural back-grounds, or other individual characteristics. The CCO can assist parents in evaluating their own needs and those of their child(ren) and then assist them in evaluating the capacity of various programs to meet these needs. The flexibility of the CCO's service alternatives provides a unique approach to supporting family value systems and parent involvement in day care arrangements.

The CCO takes a strong orientation toward the local community and allows the community's needs to shape its development and direction. A CCO in another community would, undoubtedly, look quite different from ours. Single industries or an entire community could adopt this model to fit their needs. To avoid duplication of efforts and services, it should also be possible for industries and community agencies to cooperate to develop one CCO to serve the entire community.

BIBLIOGRAPHY

Barnard, Amelia B., ed. *Community Services in Metropolitan Denver.* Denver: Mile High United Way, 1974.

Breitbart, Vicki. *The Day Care Book.* New York: Knopf, 1974.

Caldwell, Bettye M. "What Happens to Children in Day Care?" Paper presented as an S. and H. Foundation Lecture at Pacific Oaks College, Pasadena, Feb. 9, 1972.

Caplan, G. *Theory and Practice of Mental and Health Consultation.* New York: Basic Books, 1970.

Fein, Greta G. *Day Care in Context.* New York: Wiley, 1973.

Frankenburg, William K. and Duncan, Burris. *Colorado Directory of Services for Children.* Denver: 1971.

Gerald, Patricia. "The Three Faces of Day Care" in *The Future of the Family,* ed. by Louise Kapp Howe. New York: Simon and Schuster, 1972. Pp. 268–282.

Handler, Ellen. "Expectations of Day Care Parents." *Social Service Review* 47 (June, 1973): 266–277.

Hoffman, Gertrude L. "The Day Care World of Children." *The Woman Physician* 26 (Mar., 1971): 151.

Larrabee, Margery M. "Involving Parents in Their Children's Day Care Experiences." *Children* (July–Aug., 1969): 149–154.

Long, N. "Information and Referral Services: A Short History and Some Recommendations." *Social Service Review* 47 (Mar. 1973): 49–62.

Mass, Sidney Z. "Integration of the Family into the Child Placement Process." *Children* 15 (Nov. 1968): 219–224.

Prescott, Elizabeth and Jones, Elizabeth. "Day Care for Children: Assets and Liabilities." *Children* 18 (Mar.–Apr., 1971): 54–58.

Ruderman, Florence A. *Child Care and Working Mothers.* New York: Child Welfare League of America, Inc., 1968.

Wallach, Lorraine B. and Piers, Maria W. "Family Day Care: The Humanistic Side." *Child Welfare* 52 (July, 1973): 431–435.

Winter, Metta L. and Peters, Donald L. "Day Care Is a Human System." *Child Care Quarterly* 3 (Fall, 1974): 166–176.

Zamoff, Richard B. and Lyle, Jerolyn R. "Who Needs What Kind of Day Care Center." *Child Welfare* 52 (June, 1973): 351.

Chapter Seven

A PLAN FOR THE HEALTH CARE OF CHILDREN

Ann DeHuff Peters, M.D.

THE PROBLEM

Perhaps the single most consistent feature of health-care delivery for children in the United States has been its inconsistency. Whereas authorities in the fields of pediatrics and maternal and child health have long advocated "consistency, continuity, and comprehensiveness" (1) of health care for young children, the most vulnerable age group in mortality and morbidity except for old age, actual development has been almost the exact opposite—fragmentation, discontinuity, and episodic crisis care. Whether the argument that consistency, continuity, and comprehensiveness is practiable or even ideal is ever settled (2,3), the facts remain that the delivery of health services is still markedly skewed in the direction of the upper socioeconomic strata (4), and that the use of allied health professionals to aid in the reorganization of health-care delivery to reach more children is proceeding slowly (5).

As a pediatrician and public health physician, I have long advocated the development of health-care programs within day-care settings as a means of providing care to many young children hitherto neglected (6,7). In an experimental research demonstration program at the Frank Porter Graham Child Development Center of the University of North Carolina, Chapel Hill, the provision of direct health care to the children was an essential part of the program from its inception in 1966. A number of studies by the health team indicated that such a plan was not only practical in terms of acceptance by children, parents, staff, and community (8), but even more important, posed no added risk of infectious disease transmission even when ill children were routinely admitted along with those who appeared healthy (9).

Long-established beliefs die hard, however, and it is still the practice in most of the United States—indeed in other countries as well (10)—to exclude ill children from licensed group day-care settings (11). Few people recognize that continuity of relationships and settings, an important factor in the emotional well-being of young children, is threatened by this practice. In addition, wherever mothers are required to stay home from work to care for ill children without additional paid leave, financial and other family problems result almost without exception. A recent study for the Department of Health, Education and Welfare (12) reports that employer-subsidized child care has not improved absenteeism nor employee turnover. The obvious answer is that maternal absenteeism, job mobility, and welfare dependency for many mothers will continue as long as neither employer nor child-care program looks at all aspects of the needs of child and family. A major problem in maternal absenteeism is illness of the child. As long as child-care settings continue to exclude the ill child, this problem will continue.

An added problem has been the lack of attention to the mental health component of child-care programs, both in regard to child and family and to the child-adult relationships in the day-care setting. Mental health and physical health cannot be viewed in separate, categorical fashion. They are closely interwoven, most particularly when very young children and young parents are involved. This country has made little attempt to design good, practical, effective ways to help young parents with their child-rearing tasks. Indeed, in my opinion, we as professionals have done everything we can to discourage parents from receiving appropriate help with this important responsibility, and have then criticized them severely for not seeking us out. We hide ourselves in offices, clinics, and schools, keeping office hours only during daytime working hours, and surround ourselves with thousands of barriers, both personal and financial, that insulate us from young families. We need to devote some careful thought to making our services more useful and effective to those who need it most.

Parents of all socioeconomic levels and the staffs in most child-care settings show a disturbing ignorance of all aspects of health. In only a few instances have training programs for child-care staff included any up-to-date content about health: preventive health care, nutrition, interpretation of behavior, observation of child-adult interaction, recognition of symptoms of physical illness, treatment of emergencies, transmission of infection, and simple personal practices such as handwashing and prompt discarding of used paper tissues. As knowledge and practice in pediatrics and child psychiatry has risen in sophistication, with resulting specialization, the communication of simple, ordinary, practical information to parents, to child-care givers in all settings, and even to other professionals such as teachers and child welfare specialists has lagged far behind.

Day Care of Children: Organization or Disorganization?

Day care for children in the United States has grown like Topsy, with little community planning or design and with relatively little community understanding of its potential value as a family support service. Studies and surveys cited elsewhere in this volume show wide variation in type, quality, and coverage of organized day-care services. The National Children's Bureau (13) has estimated that 90 per cent or more of children receiving daytime out-of-home care are in private family homes, either with relatives, neighbors, or others less closely identified with the family. Only a small percentage of these family day-home arrangements are licensed or receive any type of outside assistance with the complexities of providing care for other people's children. The picture in group care is not much better. In contrast with European countries, where child day care has been developed under planned auspices, it is possible for anyone in the United States to open a child-care program or to take children into the home, provided he or she meets the licensing requirements of the state agency, which vary widely from state to state. In addition, numerous other programs do not have to be licensed, such as those under the direction of other public agencies, federal programs such as Head Start, and those on military posts.

In a recent study of child-care services in one large California city, I found at least six different types of group child-care programs, in addition to the countless varieties of formal and informal family day care arrangements. The group care settings range from small (10 to 75 children) to large (100 to 250 children), from private owner-operated to public-supported, and from "developmental" to less than "custodial" in nature (14). Licensing and regulation of group programs are carried out by a state licensing staff responsible for all kinds of group programs (not just child day care), while the licensing and supervision of family day-homes is the responsibility of an entirely different staff in the county department of public welfare responsible also for the licensing of all kinds of "out-of-home" family care.

Education of the "caregivers" and directors of programs and of family day-mothers varies from master's degrees in early childhood education to "certification" in elementary education, to a high school diploma or less. There is no thread that weaves centers and homes together; indeed very little thread of communication links group programs themselves, and there is no readily available source of technical assistance for anyone: program operator, staff member, family day-mother, parent seeking help, or individual wishing employment. Indeed, the situation exemplifies the chaotic conditions described so vividly by Mary Keyserling in her report "Windows on Day Care" (15).

With regard to health, the picture is the same. Only one type of group care program other than the three Head Start centers offering full day care has any formal health care organization. This program, the Children's Centers of the Unified School District (financed by state funds and Federal Title IV-A monies but administered by the local school district), has limited service from the school nurses and physicians, who function in a traditional type of school health program. In an effort to improve existing services, the supervisor of school nurses has obtained training for several of her nurses through a short course in physical diagnosis developed especially for them by one of the community pediatric centers. One of these school nurse-practitioners has been assigned full time to those children's centers located in the low income census tracts of the city. Since children with any type of symptomatic illness are excluded from attendance, she does not provide any primary health care. Her duties consist chiefly of spot-checking for acute illness, identification of potential problems, and counseling with staff and parents. She has limited medical backup, since the school physicians are concerned largely with accidents and other problems of crisis in the school setting and the responsibility for continuing health care of the children is in the hands of the "physician of the family's choice." Covering several centers, her time with each is limited.

In this same community, there is no source of organized health consultation for day-care programs, nor any provision for care of ill children such as the Berkeley Home Health Aide Program (see p. 75). Health content in training programs is meager, and only in a few instances has a physician been involved in any aspect of such instruction. The slight amount of health-oriented information available to a few individuals in child-care programs comes from Red Cross nursing personnel.

Some provocative studies now in progress under the direction of Dr. Charles Lewis at the School of Public Health, University of California at Los Angeles(16) indicate that personal behavior and attitudes about health and health care are determined long before the age of entrance into public school. This early conditioning may have both positive and negative effects, according to Lewis and his staff. Possibly positive effects include acceptance of simple health precautions, understanding the value of a balanced diet, and seeking help when symptoms such as fever, continuing malaise, and severe cough occur. Among negative effects are excessive concentration on bodily symptoms of all kinds, making frequent contact for minor problems with health professionals such as the school nurse, and fears related to bodily functions. The population studied by Lewis came from the same city described above, which has little or no up-to-date health-oriented training for staff in preschool programs. Since the study group consisted of children in elementary school, it would be most informative to move downward to even younger children in an attempt to find the fine line that distinguishes overconditioning from appropriate health-care-seeking behavior. An inevitable question results. As we in this country, by our hypocritical approach of advocating early and continuing health care but providing it only for a privileged few, contributing to increased handicaps of all sorts, ranging from the defects that result from inadequate medical care to a pathologic dependency upon frequent medical contacts?

ATTEMPTS TO SOLVE THE PROBLEMS

Since the economic organization of this country is not oriented toward the provision of adequate maternal leave to allow working mothers to attend to health needs of children, the responsibility for deciding how best to protect and enhance the health of children has by default been left to the day-care providers. Gradually, health professionals also are beginning to look at these issues and are trying to help find solutions. There are three issues: how to help families obtain necessary health care for their children, how to teach the care-giving adults appropriate health-oriented behavior, and what to do with the child who is sick.

Studying the first question, we find a wide variation in organization and practice over the United States. The fiction that all families have ready access to reliable sources of primary health care still exists in a few sections of the country. In a few other areas, public health departments have expanded their services beyond the traditional clinics for immunization and periodic well-child evaluations to provide public health nursing visits to publicly sponsored child-care programs. In other parts of the United States where broader health coverage has been attempted through public health departments, child health services are made available directly to children in the day-care setting. For example, the Guilford County Department of Public Health in North Carolina has equipped a mobile examining unit as part of its children and youth project, and sends the unit to day-care centers for periodic examination and treatment of the children. It can, however, provide service only to a limited number of children, since it also has other responsibilities under the children and youth program. Another health department, in Maricopa County, Arizona, includes day-care programs in its health service coverage, with an emphasis on providing preventive health services of all kinds to children in centers. In the Morrisania health district in the borough of the Bronx in New York City, plans are under way to provide health care for the children in five community day-care centers (and their siblings up to age 18) by utilizing the services of the Montefiore-Morrisania comprehensive child-care project (a children and youth program of the New York City Health Department and Albert Einstein College of Medicine). This program will provide all types of health services—preventive, acute episodic care, and diagnosis and treatment of long-term problems.

None of these programs reach children in family dayhomes, unless the family day mother or the child's own parents is aware of the service and initiates contact. In addition, these plans resolve only a portion of the health care problem, since they do not include health education of child-care staff, nor provisions for the care of ill children. A number of communities have been examining ways to help parents with children when they are sick. In Berkeley, California, a home health aide program, financed by a grant from the City Council, provides a central switchboard and ten to twelve trained home health aides who go into the homes of ill children to care for them. It is publicized by radio and newspaper. A small fee is charged for the service on a sliding scale, with the maximum about $2.50 an hour although the actual cost is figured at about $4.00 an hour.

The aides, needless to say, are kept busy, and the demand is more than can be met. Other communities have considered plans to open a type of infirmary where ill children can be cared for; but, although I have received numerous requests for help in development of this type of plan, I am not aware of any that are as yet in operation.

There are several problems inherent in such a plan: (1) difficulty in defining acute illness in young children in terms of type, duration, and etiology; (2) problems of mixing children from different programs, often with different kinds of acute infection; (3) the lack of emotional support and stability in such arrangements at a time when a child most needs the familiarity and comfort of known and loving adults; and (4) the increased incidence of staff illness in caring for children with infectious illness. Studies at the Frank Porter Graham Center previously cited, where children attended every day whether sick or well, demonstrated that certain diseases are more likely to affect children at certain ages. For example, during a community epidemic of mumps, no child under age two in the center showed any evidence of virus infection, although all of those over two developed mumps virus antibodies whether or not they had physical symptoms of the disease. The long-term studies of antibody formation showed that certain microorganisms appeared to infect more of the three to four-year-old children than the younger ones, for example, and symptoms differed in some respiratory infections according to the child's age.

Caring for ill children in family day homes seems to pose fewer problems, although data from these less visible child care settings is less reliable. The small number of children and the wider age spread within most homes would appear to simplify problems of care. However, the lack of organized health assistance to family day homes and the relatively unsophisticated level of health behavior of many day home mothers that I have witnessed might counterbalance these apparent assets.

I do not feel that plans to organize sick-care centers or infirmaries of the type mentioned previously are advisable from any standpoint of mental or physical health of child or family, and I do not recommend their implementation. They are artifical, expensive, and hazardous. In countries such as the Union of Soviet Socialist Republics, such separate facilities are provided for ill children who are in day-care centers, but they are adequately staffed, controlled, and financed in a way that would be impossible under our free-enterprise haphazard service "nonsystem." If it is impossible for a mother to stay home with her ill child or for care to be provided in the child's familiar day-care setting, whether it be family day home or day-care center, it would seem far more advisable to train and use the previously mentioned home health aides as a temporary expedient.

The basic issue, therefore, is one of education: education of the health professional in the needs of children in day care, education of the parent in knowing how and where to secure help, education of the care-giving adults in knowing what to look for and how to make decisions, and ultimately, education of the children themselves in how to assess their own physical symptoms as they grow older.

A PLAN TO TIE TOGETHER EXISTING HEALTH SERVICES TO REACH CHILDREN IN DAY CARE

A comprehensive health program covering all children in day-care settings lies beyond the reach of service-delivery systems currently available in this country. Whatever the eventual desirability of such a broadly scaled program, planners should consider it as a likely improvement over the fragmentary, episodic health care that children now receive. As an alternative, some method should be sought to weave a thread of continuity in presently existing services. Such an attempt could not only identify and make use of available health programs and resources, perhaps increasing their effectiveness, but would also aid in pinpointing gaps and unmet needs. One of its major aims would be community education in health aspects of child care.

The following plan would apply primarily to metropolitan communities. However, as experience is gained, modifications or adaptations could be devised to provide similar services to rural areas or to smaller points of larger urban settings. The plan consists of developing a health consultation team, administratively attached either to the local department of public health or to a medical teaching institution, or both. The team would function in the field —in day-care programs, family day homes, and health-care settings—as its members visit those requesting help, those referred for special service, or those providing health care.

The team would consist of the following members: a physician, preferably a pediatrician, with interest and experience in broad aspects of child development and an understanding of the needs of families and children in the day-care settings, a nurse-practitioner or a public health nurse with training and experience in child development, and one or more community health aides. The latter might be licensed practical (vocational) nurses or individuals trained by some other type of formal course work and practical experience with emphasis on family and child health.

Responsibilities of this team would include the following:

> *Direct service* to child-care centers and family day homes in such matters as development of health policies, organization of health in-service training for staff of a center or a group of family day mothers, contact and planning with physicians providing primary care to

children in the day-care settings, counseling with individual parents as indicated or desired, aid with referral to sources of care not previously known to family or program, arrangement for special services or evaluations, and assistance with followup or referrals.

Development of training components for formal course work in educational institutions in broad health aspects of child development, problems of mental and physical health in early childhood, evaluation of symptoms of acute illness in children, organization of health care for children and staff, parent couseling in health for child-care staff, and other related subjects.

Development of public information materials; conferences with medical, nursing, educational, and social work groups; organization of a health-resource library for child-care givers and parents; and participation in community organization for all types of child-care settings and all aspects of child-care development.

My experience with multidisciplinary programs has shown that defined roles often change as programs progress. Therefore, I shall not outline specific details and priorities of how the major components of the consultation service should develop. But I would like to emphasize three areas.

The Physician

I feel a physician is an essential part of coordinating such a health consultation service. A physician performing in the usual way—sitting in an office looking over records and making recommendations from a "paper study" or making episodic contacts on a crisis basis—may indeed make poor use of highly trained and high-priced medical time. However, a physician out in the community visiting programs and health providers, talking with community groups, and helping to put together a viable and evolving consultation service with some continuity and permanence, could make valuable use of skilled medical time. In the latter circumstance, "diagnosis" and "treatment" as applied by McGavran to a community rather than to an individual patient (18) would appear to be appropriate use of physician time. Also, whether we like it or not, it is still a cultural fact of life that words spoken by a physician nearly always carry more weight with other physicians, especially with regard to program planning.

The Nurse

The nurse, by training and experience uniquely equipped as a family counselor, should be given responsibility for as much primary care as the system and her time will allow. In some sections of the country, for example, she might provide direct primary care for children in some programs. In other areas, she could more appropriately work through other nursing systems such as school nurses or public health nursing staffs. She also should have opportunity to work directly with community physicians in plans for individual children, since many busy practitioners welcome such assistance. As she demonstrates what good primary care given by a nurse can be, she may generate support for additional development of other similar primary health-care programs in the community (see Chow's article on providing health services).

The Health Aides

The community health aides undoubtedly will need education for their new roles, since few health training programs designed for individuals working in child-care settings now exist in the United States. Such education will have to be given by the physician and nurse members of the consultation team. As this proceeds, physician and nurse will also learn. These trained aides spend blocks of time in day-care settings, identifying problems of daily practice, demonstrating good care-giving techniques, especially for infants and toddlers in those programs courageous enough to take ill children, and helping with staff in-service training. A projection of such use of licensed practical nurses, and my experience with these community people who understand much that professionally trained people overlook, suggest that such a plan would not be expensive and could pay long-term dividends in better daily care of children and more appropriate use of all types of health professionals. A project for family day mothers recently completed by Sale and her co-workers at Pacific Oaks College (19) found these women eager for help and support of all kinds in their tasks of rearing children from other families. The family day mothers welcomed the regular visits of students to their homes and developed such a strong feeling of comradeship that they continued to meet even after the research project had come to an end. Community health aides operating as members of a health consultation team could provide similar assistance to family day mothers and might significantly raise the efficacy of heath-oriented performance both in the day-care families and in the families of the children themselves.

I have had experience during the past three years in some of the responsibilities outlined previously, first as a staff member of a complex of privately financed child health services, and lately as an independent consultant. I am convinced that such a community health consultation service would be not only feasible, and relatively inexpensive in total dollars spent per child or family, but welcomed by both child-care and health-care providers.

NOTES

1. Concepts expressed in the Conference on Health Services for Children and Youth, Chapel Hill, N.C., March 18–20, 1969. *Amer. Jour. Pub. Health* 60 (April, 1970), Supplement. Also refer to Wallace, Helen M. *et al.* "Comprehensive Health Care of Children." *Amer. Jour. Pub. Health* 58 (Oct., 1968): 1839–1847.

2. Gordis, Leon and Markowitz, Milton. "Evaluation of the Effectiveness of Comprehensive and Continuous Pediatric Care." *Pediatrics* 48 (Nov., 1971): 766–776.

3. Haggarty, Robert J. *et al.* "Symposium: Does Comprehensive Care Make a Difference?" *Amer. Jour. Dis. Child.* 122 (Dec., 1971): 467–482.

4. Of the numerous studies made during the 1950s and 1960s, two representative ones are: Peters, Ann DeHuff and Chase, Charles L. "Patterns of Health Care in Infancy in a Rural Southern County." *Amer. Jour. Pub. Health* 57 (Mar., 1967): 409–423, and Mindlin, Rowland L. and Densen, Paul M. "Medical Care of Urban Infants: Health Supervision." *Amer. Jour. Pub. Health* 61 (April, 1971): 687–697.

5. Egeberg, Roger O. and Pesch, Leroy A. "The Role for Nurses." Editorial. *J.A.M.A.* 220 (May 29, 1972): 1237–1238.

6. Peters, Ann DeHuff. "Health Support in Day Care." In *Day Care: Resources for Decisions,* ed. by Edith Grotberg. Chapter 11. OEO Pamphlet 6101–1. Washington: U.S. Government Printing Office, 1971.

7. Peters, Ann DeHuff. "The Delivery of Health and Social Services to Child and Family in a Daytime Program. In *Early Childhood Development Programs and Services: Planning for Action,* ed. by Dennis N. McFadden. Columbus: Battelle Memorial Institute, 1972.

8. Loda, Frank A. "Group Care for Children under 3: Experience with a Program Providing Primary Health Care." Paper presented at the Annual Meeting of the American Public Health Association, Oct. 28, 1970.

9. Loda, Frank A.; Glezen, W. Paul; and Clyde, Wallace A., Jr. "Respiratory Disease in Group Day Care." *Pediatrics.* 49 (March, 1972): 428–437.

10. Davidson, F. *et al. Care of Children in Day Centres.* Public Health Paper No. 24. Geneva: World Health Organization, 1964.

11. Social and Administrative Services and Systems Association and the Consulting Services Corporation of Seattle. *Compilation of State Regulations for Day Care Centers and Group Day Care Homes.* Seattle: National

Day Care Licensing Study of the Office of Child Development, 1971.

12. Ogilvie, Donald G. *Employer-Subsidized Child Care.* Washington: A report prepared for the Department of Health, Education and Welfare by the Inner City Fund, 1972.

13. Low, Seth and Spindler, Pearl G. *Child Care Arrangements of Working Mothers in the United States.* Children's Bureau Publication No. 461. Washington: U.S. Government Printing Office, 1968.

14. Day Care and Child Development Council of America, Inc. *Standards and Costs for Day Care: A Compilation.* Washington: (Mimeo.) DCCCDA, 1970.

15. Keyserling, Mary Dublin. *Windows on Day Care.* New York: National Council of Jewish Women, 1972.

16. Lewis, Charles S. Personal communication.

17. Hoekelman, Robert A. "A 1969 Head Start Medical Program." *J.A.M.A.* 219 (Feb. 7, 1972): 730–733.

18. McGavran, Edward G. "The Challenge of Public Health to the Medical Man." *Texas State Jour. Med.* 54 (Sept., 1958): 634–639.

19. Sale, June S. *et al. Open the Door . . . See the people. A Descriptive Report on the Second Year of the Community Family Day Care Project.* Pasadena: Pacific Oaks College, 1972.

Chapter Eight

HEALTH CONSULTATION
IN CHILD CARE

Susan S. Aronson, M.D.

This paper was written to give the reader a broad overview on the who, what, how, and why of health consultation. It covers the following topics:

Why is a health consultant needed?

What does the "health component" of the child care program include?

What is special about the consultant-client relationship?

How do you get the most from your consultant?

How do you find a consultant?

How do you make sure that the health component improves after the consultant gives his advice?

What can be learned from the mistakes others have made in using a health consultant?

WHY IS A HEALTH CONSULTANT NEEDED?

The clients who seek health consultation may represent any program type: private or public, full- or part-day, center or home based, developmental or custodial child care. Frequently the significance of the program type is overstated by the client, who is accustomed to defending the uniqueness of his service to parents and others. The basic common objective in any program's approach is meeting the needs of the child and his family. This objective is common to all types of child-care programs. The type of program does affect the methods and extent to which family needs are addressed, but the program type does not really change family needs. In fact, the most useful thing the consultant may do is to help the program see that its type of approach cannot meet the needs of all families. The consultant can help the program to examine its strengths and help devise ways for the program to identify and assist families who are not well served to find another service which can do a better job for them.

WHAT DOES THE "HEALTH COMPONENT" OF THE CHILD-CARE PROGRAM INCLUDE?

The health component is one of several aspects of child care. Other aspects include education, social service, parent involvement, and administration. A global definition of the health component includes (1) preventive and remedial medical services, (2) nutrition services, (3) psychological services, and (4) dental services; plus health education and administrative practices in each of these four areas. The health consultant can use his expertise to help the program establish goals and evaluative measures for each area of the health component, but it will require the program's expertise to work out meaningful objectives and reasonable procedures to accomplish the goals.

Health services usually are provided by health resources outside the child-care program, but parents must be encouraged and aided to secure these services for their children. We all know how hard it is for working families to bring their children to the health provider for routine care, but in the larger view it is very important to help each of the child-program families to identify a health center in which they can obtain direct medical attention. The child-care program should consider itself a matchmaker between existing community health services and the families served by the program. Unless examination, evaluation, or treatment services are completely unavailable in the community the child care program should not provide these direct health services. At face value, this recommendation might not seem to be in the best interest of families. Child-care program health services rarely can be available 24 hours a day, provide service to all family members, or continue to be available when the child no longer is enrolled in the program. While it is sometimes necessary for the program to provide some limited services in order to obtain entrance into or support a continuing relationship of the family with a health service provider, it is best not to encourage family dependency on the child-care program as the primary source for extensive health care services.

A primary consideration to program families is not to encourage the use of multiple providers for health care. In contrast with direct health services, many other health-related activities such as developmental assessment, management of minor illnesses, and daily health routines should be integrated into the child-care program. Because of day-by-day contacts with children and families, the child-care program is in an excellent position to promote, protect, maintain, and improve the health of the children and families it serves by providing such services.

Child-care programs are designed to supplement family resources and to heighten parental awareness of opportunities to meet their children's needs. Even the most loosely structured cooperative nursery can easily integrate some health concerns into the daily program. Many health activities require a large amount of increased awareness to only a small amount of additional effort on the part of the child-care program. The goal of the child-care program health component should be primary advocacy for health. This goal can be addressed by (1) fostering improved health habits among the children, parents, volunteers, and community; (2) considering safety and environmental quality along with other issues in planning the daily activities which make up the program; (3) seeking opportunities to act as a liaison to assure appropriate utilization of health service resources; and (4) seeking the advice of health professionals in health component planning.

Most of the issues in health will overlap with other program aspects. This overlapping is sometimes difficult for child-care program staff to accept. Social service, education, psychological, and administrative professionals may need to be reminded that few children are divided into sections which match our professional labels. We need to work together to find the best solutions to help the child and his family. Few problems can be solved by labeling them. As an example, a child who has frequent earaches has a health problem, but because of his illnesses, his parent loses sleep, performs poorly at work or school, and may be unsympathetic to the emotional needs of the child. The child himself may not progress well developmentally because he does not feel well. Is this a health problem, a social service problem, a psychological problem or an educational problem? Who should work with the family? By addressing the many facets of such problems openly, the health consultant can assist the child care program in evolving real team work.

The health consultant can best help the child-care program if he or she is well informed about the realistic limits imposed by budget and competing priorities of the child-care program. The better informed he is, the better he can use his ingenuity and his access to additional sources of expertise to propose approaches to solving health related problems. Early in the health consultation process, the health consultant and child-care program staff should critically evaluate the program's accomplishments in the various areas of the health component. For each area of the health component the following questions should be asked:

1. What are the needs of the children, staff, and parents in this area? What is the program presently doing to meet the needs?

2. Who is involved in this aspect of the health component now? Who should be involved? (Remember to include those affected by the activity, those with authority over the activity, and those with expertise to bring to bear on the activity in this type of planning).

3. How is this aspect of the health component being addressed? How should it be addressed?

4. Where is this aspect of the health component taking place? Where is it best addressed?

5. When is this aspect of the health component accomplished? When should it be done?

In all areas of the health component, the consultant should help program personnel to understand clearly why the activities to be performed are important; in this way, a meaningful commitment can be made by all those involved.

It may be useful to organize this evaluation around the various areas of the health component.

Health Service: Screening, Treatment, and Follow-up

Intensity of the health service activities of the child-program depends to a certain degree upon the extent to which children and their families have easy access to health care providers. Middle-class children, too, can benefit from advocacy efforts to assure each has received appropriate preventive health services and that health concerns of program staff are brought to the attention of the health provider. Private health care providers usually do not make vigorous follow-up and preventive outreach efforts and may need to be educated about the potential support that the child-care program can give the family in complying with the health provider's recommendations.

Health Records

The child-care program can play an important liaison role in helping parents to identify available health services by maintaining with parents complete and up-to-date health records. A child-care program's health record need not duplicate the detail or sophistication of a health provider record, but it should document which services have been provided to each child, who provides health services to the child, and any services needed to ensure recommended child health standards.

Health Policies and Procedures

Written policies and procedures help staff, parents, and consultants reach a common understanding of the health component. Clearly documented procedures and policies minimize the indecisiveness as to which routine is to be followed in a given situation, the appropriate individuals to contact, and the limits of the responsibility to the parents, the program staff, and the health consultant.

Attention to Children with Special Needs

Often it is felt that children with special needs require a grossly restructured program and environment. However, the primary health care (basic health needs) of special children is not different from that of "normal" children. A good health component should be able to meet the primary health care needs of children with special needs and those without. When special needs children are in care, additional attention to staff training around the area of special need is highly desirable. The program staff often is unaware of how to deal with certain impairments experienced by children in their program. As an example, staff may be unaware that when speaking to a child who has a hearing deficiency, it is not necessary to shout or speak slowly. A brief consultation with a knowledgeable consultant can dispel commonly held misconceptions and make the integration of the child far more comfortable for both child and staff. With the current emphasis on mainstreaming of children with mental and physical handicaps, child-care programs must learn how to use existing resources to supplement the basic services already provided by the program, but not feel overwhelmed by the misperception that totally different programming is required.

Health Education

In this area of the health component, all other areas can be reasonably addressed. Health education includes emergency preparedness, knowledge of first-aid management and appropriate methods of management of minor illnesses, awareness of the various aspects of the safety and quality of the environment over which we have some control, and the rationale for any of the efforts made in the health component.

Health education includes the general education of all those who may be affected by accidents, illnesses, or common health problems. It is beneficial for everyone—children, staff, parents, volunteers—to be aware of his role in fostering health and be prepared to respond as an advocate for positive health behavior in day-by-day situations.

Staff Health

The program should assure the hiring and appropriate supervision of all individuals in the program. Unusually frequent or long absences for illness should be investigated. In addition, initial and ongoing annual screening of staff and volunteers for infectious disease problems, assessment of chronic diseases which require special regimens or medication, mental stability, and exertion tolerance should be conducted. Such screening must not violate the right to privacy of the adults involved, but it can be justified wherever there is potential impact on the quality of care provided by the program.

Dental Care and Daily Dental Hygiene

As with many areas of the health component, dental health education is best accomplished by incorporation of dental hygiene into daily routines. The simple practice of daily toothbrushing, rinsing with water after sweets, and frequent use of detergent foods in the menu planning can promote dental health throughout life.

Evacuation, Emergency, and Disaster Plans

By visibly posting and practicing emergency plans and evacuation procedures, loss of life and serious injuries may be prevented. In line with health education, each individual should know his or her role in emergencies and should perform in such a way as to increase the efficiency of others when dealing with an emergency situation.

First Aid and Management of Minor Illnesses

Program staff should know first-aid methods. A nurse is not required to detect illnesses or treat a minor injury; parents do it all the time. Preliminary steps should be taken by any staff member present when an accident occurs or when a child appears ill, even when professional health advice is desired later.

Safety and Environmental Quality

Safety and environmental quality measures have suffered from benign neglect by child-care programs. It is true that an environment cannot and should not be expected to be absolutely safe; but this does not justify ignoring the simple ways to improve the safety and quality of the environment. Preventive measures include storing cleaning agents in safe places out of the reach of children; rotating the responsibility for a routine safety survey among staff volunteers and children; installing inexpensive dehumidifiers in areas of dampness; installing portable screens where complete screening is not possible; and limiting window openings to six inches to prevent children accidentally falling out.

Nutrition

Nearly all child-care programs incorporate some aspects of nutrition into their programs by providing snacks and/or meals. By understanding and utilizing the basic elements of nutrition—sociocultural implications, nutrient quality, frequency of food and fluid intake, and economic aspects—the child-care program can have a persuasive impact on this aspect of health. Simple concepts like using the four basic food groups to promote a balanced diet can be incorporated by children and their families into life-long habits. The child-care program must model and not just talk about good nutrition.

Psychological Services

The mental health professional can be of great value to the child-care program. Although his role is frequently misunderstood or stigmatized as appropriate only for those who are severely disturbed, the mental health professional has special expertise to help staff in routine assessment of all children as well as methods to identify and plan for those with special needs. These consultants can also recommend routine procedures which will better meet the individual needs of normal children. The program staff, who work with the child daily, are often in the best position to provide information from which the mental health professional can formulate recommendations; the staff can also see that these recommendations are applied on a day-to-day basis.

Transportation Arrangements

Some means of transportation should be immediately available to the child-care program at all times. When an emergency arises there often isn't time to call a cab or rely on public transportation. Calling a parent or some other individual outside the program to provide a car may result in the loss of vital time. While police and fire rescue vehicles are commonly used, plans should include alerting these community agencies about the location and nature of the program. In addition, arrangements must be made to release a staff member to accompany the injured child without jeopardy to the supervision of the other children in the program. All vehicles used in the transportation of children and adults should comply with automotive safety standards and be equipped with seat belts. Even in programs which do not provide vehicular transport for trips or daily use, pedestrian safety plans for walking to and from the program site, nearby public transport stops, and parent education in transportation safety are appropriate.

Administrative Practices that Facilitate the Implementation of the Health Component

Each program procedure should be justified in terms of its ability to foster continuous improvement of the health component year by year. To this end, programs need (1) health records that document which services are needed and which services have been provided; (2) health policies and procedures that help all staff members, parents, and consultants to have a common understanding of the health component; (3) staff health measures that assure hiring and appropriate supervision of individuals to carry out the plans of the program; (4) posted emergency plans; and (5) appropriate accident reporting for legal and preventive reasons. Inspections and licensing procedures should be considered as opportunities for obtaining additional expertise rather than as threats to the program's continued operation. Often licensing representatives who have visited many child-care programs can suggest reasonable solutions to problems previously thought insurmountable. By focusing special attention on a particular aspect of the program, the licensing representative can identify oversights and help in the evaluation process which makes continuous improvement possible. This is particularly true of sanitarians, fire and safety inspectors, and building code representatives.

Evaluation

A convenient method of evaluating these 14 areas of the health component and assessing the need for improvement is by the use of a form called "The Health Component Improvement Plan" (Table 8-1). This form is used to (1) identify the existing problems within the health component; (2) establish a plan for dealing with the problems; and (3) establish a time schedule and priority listing for problem-solving.

In practice, the health component improvement plan has proved to be a useful tool to focus discussion of program staff and the health consultant. It can also be used for ongoing appraisal and replanning when more objective and extensive evaluation procedures cannot be used.

Integration

While we have looked at the various areas of the health component separately for purposes of this discussion, many areas frequently are united into a single activity. For example, a common program and family activity is a walk outdoors. Our environment is filled with both hazards and opportunities for risk taking. A planned walking trip provides an excellent opportunity for safety education, good modeling of appropriate street crossing and motor vehicle safety awareness, as well as an enjoyable curricular experience and a healthy exposure to outdoor air. Since preventable accidents account for the majority of health problems of children and young adults, safety considerations deserve our full attention in child care.

Generally health education for children and adults is best oriented around naturally occurring program events. The outdoor walk is one example. Another example is the occasion of a child or adult with a mild cold. Since most

Table 8-1
Health Component Improvement Plan*

Health Component Problem Area (Items in Subscale Score)	Individuals or Agencies to Be Involved in Solving Problem (Affected, Expert, Authority)	Priority Number	How Changes Can Be Made	Time Plan
1. Health Service (Including screening, treatment & follow-up)				
2. Health Records				
3. Health Policies and Procedures				
4. Attention to Children with Special Needs				
5. Health Education				
6. Staff Health				
7. Dental Care and Daily Dental Hygiene				
8. Evacuation & Emergency & Disaster Plans				

Table 8-1 (Continued)

Health Component Problem Area (Items in Subscale Score)	Individuals or Agencies to Be Involved in Solving Problem (Affected, Expert, Authority)	Priority Number	How Changes Can Be Made	Time Plan
9. First Aid and Management of Minor Illnesses				
10. Safety & Environmental Quality				
11. Nutrition				
12. Psychological Services				
13. Transportation Arrangements				
14. Administrative Practices—Internal Planning				
15. Administrative Practices—External Planning				

*The Health Component Improvement Plan is used as follows: (1) First name the problems in each area to arrive at a global picture of potential areas of improvement. (2) Next, identify who is involved in the solution and (3) how the solution is to be approached. (4) This information will allow a time line to be set for each problem. (5) Finally, using all the information on the nature of the problems and on who, how, and how long it will take to solve the problem, a priority listing can be made.

preschool children suffer seven to ten colds each year, it is worthwhile to plan for the additional rest, fluids, and personal hygiene with parents and staff to make these measures routine.

What Is Special about the Consultant-Client Relationship?

The client must want advice for the advice to have any impact on the program. Often a request for consultation comes for reasons other than those initially stated to the consultant by the client. Surfacing the real and underlying reasons for the request for consultation must be considered an important part of the consultant's and the client's responsibility to one another. Sometimes the reason for the request is that a monitoring agency has found the program out of compliance with existing regulations or codes. If this is the case, then the priorities and time frame for the consultant's advice are quite different from those of a program engaged in long-range planning. Occasionally, the underlying reason for seeking consultation is an immediate health problem such as a serious illness or accident to one of the program participants. In such a case, part of the consultant's role is to help alleviate the guilt burden of the clients and broaden the narrow focus of the program before meaningful health component improvement can take place. Events like these, however, provide for continuing contact between the consultant and client. These contacts reinforce a mutual trust which can grow into a long-term, fulfilling, and beneficial relationship between the two.

How Do You Get the Most from Your Consultant?

The success of health consultation is often threatened by the limited experience which either the consultant or the client has with this particular relationship. Few health consultants or child-care program staffs are well acquainted with the breadth as well as the limitations of health aspects of the child care. Often they both view health in child care as limited to assuring that all children have received their immunizations. Many important and easily implemented preventive measures such as routine safety surveillance and health habit education are neglected because of such a limited view. Suggestions for improving the health component are often met with a "but we don't have a nurse to do it" approach. Nurses and other health providers give lots of advice and instructions to parents, much of which is never carried out. As a minimum, child-care programs can help parents to follow through with health care instructions. This can happen only if the child-care program works out mechanisms for knowing about such instructions. If parents can be given the responsibility for carrying out health care recommendations, child-care programs do not require a nurse to act in the parent's stead.

How Do You Find a Consultant?

If you are convinced that a health consultant can be helpful, how do you find one? The most likely place to look is among the usual health care providers for the children in your program. This might include private physicians, clinics, hospitals, medical society groups, nursing associations, school health departments, or other health related divisions of local, state, or regional government. Whether you are directly paying your consultant with program dollars, or indirectly paying him by selecting a tax-paid public servant, the characteristics of the individual who is selected for this role are more important than his title. While no formal shopper's guide to health consultants is yet available, potential health consultants can be identified by their willingness to participate in community projects, by their openness with their consumers, and by their flexibility in face of new problems.

If your health consultant is a provider of health services, most of his experience and training has been in dealing with health concerns of individuals on a one-to-one basis. For this reason, the child-care program may need to orient the health consultant with the nature of the child-care program, with locally available resources outside of health, and with the regulations and current constraints under which the program operates. Often the health consultant is completely unaware of internal priorities which have an impact on the implementation of his recommendations. His lack of awareness can be a serious barrier to a meaningful relationship between the health consultant and the child-care program personnel. Health professionals are accustomed to being prescriptive and directive around health issues and may need assistance to develop new group work skills. Health professionals are often expected to lead the group as well as supply health expertise. This expectation may lead to a very unhappy experience for program and consultant. By recognizing that consultation and group work skills are not part of health professional training, the child-care program may be able to provide the group direction to help the health consultant feel comfortable outside his usual environment of office, clinic, or hospital. Clearly stated program expectations can prevent the foot-in-mouth disease which so commonly affects health consultants.

How Do You Make Sure that the Health Component Improves after the Consultant Gives His Advice?

After the health consultant and the child-care program have understood and agreed upon the health component and those portions of it which they choose to address, it is helpful for a single individual to be responsible for the overall implementation of any decisions or plans made. This designated individual's relationship to the other staff, to policy makers, and to consultants should be made clear to everyone, so that problems will be directed to his attention and he will be provided opportunities to assure that the program's plans are carried out. This individual's role is to assure that the plans made by decision makers or a planning body are carried out at the particular site of service on a day-by-day basis; he can most appropriately be called the health advocate. In a large program, a health advocate's sole role may be to oversee the implementation of the health component of the program. In smaller programs, the health advocate is likely to have other duties; but the program will recognize him as the most logical individual to assure appropriate integration of health activities with other activities of the child-care program. Unless one individual is stipulated as responsible for health advocacy, it is unlikely that even the best laid plans will be carried out.

The health advocate should be someone who is available to the program on a daily basis so that he can be the gadfly for fostering health in the child-care program. At the same time that he is maintaining the health component for his program, the advocate must be able to yield to the other priorities of families and of other child-care program staff. Because he is involved with daily activities of the program, he will be able to propose compromises and alternative solutions that might not be thought of by any outside person.

The person chosen to be the health advocate should also be someone who is willing to be aware of his own personal health prejudices. It is all too common for people to believe firmly in a particular approach to a health problem without being able to defend their belief. There are many such areas in health that are associated with cultural or personal preferences. How long and how often should a child play outdoors in cold weather? Should children with minor problems such as colds or teething be kept at home and excluded from child-care programs? Are such children better off at home than in the child-care program if home care means care by unfamiliar substitute caregivers because the parents must work or attend school? With an average incidence of seven to ten upper respiratory infections per child per year (1), no program will truly succeed in excluding all ill children all the time. We now know well that such children need not be excluded to protect the others, since they are most infectious before any symptoms are visible (2). The health advocate can help the program focus on this question: What special provisions can be made for meeting the needs of the ill children in this particular program today, with today's staffing?

The health advocate may delegate various aspects of the health component to other members of the child-care program staff, to volunteers, or to outside agencies. However, the essential role of the health advocate is to ensure that all aspects of the health component are attended to and are integrated with other program aspects into the overall program. He ensures that the pertinent health concerns of the staff, parents, and volunteers are brought to the attention of the appropriate health consultants.

Health advocates can be caregivers, non-health professionals, or administrators with special interest in the health area. Since in most programs health will not be the advocate's sole focus, he must be assisted by his supervisory staff and co-workers to identify the best time and circumstances in which he can perform his role. The health consultant must be advised about the necessary limited availability of the health advocate so that he, too, does not intrude upon the advocate's attention to his other responsibilities.

What Can Be Learned from the Mistakes Others Have Made in Using a Health Consultant?

It is quite common for the health consultant to be used wrongly. In many programs, for example, not one but several consultants address the specific problem area. While this is reasonable as a means of obtaining a variety of views on a particular subject, this practice can result in counterproductive recommendations. It is best to let all consultants know about other opinions and other sources of information that have been sought by the program. This enables the consultant to focus his answers as reasonable courses of action and to clarify the reasons why his recommendations may differ from those of another consultant.

Another type of misuse occurs if a consultant uses his entrée into a program to solicit requests for paid services not included in his consultant role. Unless it needs the particular services offered, the program should consider itself under no obligation to purchase services directly from the consultant. On the other hand, a program may be tempted to try to obtain direct services from a consultant which should most appropriately be paid for. By abusing a consultant in this way, the program may in fact discourage his continued availability to the program as a consultant and make him more reluctant to offer suggestions or service in the future.

Notes

1. Loda, Frank A., Glezen, W. Paul, and Clyde, Wallace. "Respiratory Disease in Group Day Care." *Pediatrics* 49 (March, 1972): 428–437.
2. Gellis, Sidney. *Yearbook of Pediatrics,* Chicago: Yearbook Publications, 1972.

BIBLIOGRAPHY

Aronson, Susan S. *The Health Component of Child Care.* Child Welfare League, in press.

Bank Street College Day Care Consultation Service. *Final Report.* 1974.

Department of Health, Education, and Welfare. *Guide for Day Care Licensing.* Washington: 1972 (in revision).

———. *Nutrition and Feeding of Infants and Children Under Three in Group Day Care.* Washington: DHEW Publication No. (HSM) 73-5606, 1971.

Friedman, David. *The Health-Professional as a Health and Mental Health Consultant: How to Be an Effective Consultant to Child Care Programs.* American Academy of Pediatrics, Subcommittee on Health and Mental Health Consultation of the Committee on the Infant and Pre-School Child, 1971–73.

"Head Start Program Performance Standards." *OCD-HS Head Start Policy Manual.* Washington: 1973.

Loda, Frank A., Glezen, W. Paul, and Clyde, Wallace. "Respiratory Disease in Group Day Care." *Pediatrics* 49 (March, 1972): 428–437.

North, A. Frederick. *Health Services: A Guide for Project Directors and Health Personnel.* Washington: Office of Child Development, 1971.

"Part 220-Service Programs for Families and Children: Title IV Parts A and B of Social Security Act." *Federal Register Volume 34.* Number 18. January 28, 1969.

Peters, Ann DeHuff. "Health Support in Day Care." In *Resources for Decisions,* ed. by Edith Grotberg. Washington: Office of Economic Opportunity, 1971.

"Proposed Regulations for Child Care Programs in Pennsylvania." Unpublished draft presented to the Pennsylvania Department of Public Welfare. August, 1974.

"Recommendations for Day Care Centers for Infants and Children." *American Academy of Pediatrics,* 1973

Richmand, Julius B. "Pediatric Aspects of Day Care and Institutional Care." *Care of Children in Day Centers.* Public Health Papers No. 24. Geneva: World Health Organization, 1964.

Schonfield, Hyman K., Heston, Jean F., and Falk, Isidore S. "Number of Physicians Required for Primary Care." *New Eng. Jour. Med.* 286 (March 16, 1972): 571–576.

Somers, Anne R. "Health Care in Transition: Directions for the Future" *Hospital Research & Educational Trust.* Chicago: 1971. Chapters 5 and 6.

Chapter Nine

PLANNING HEALTH SERVICES
A CHILD-CARE CENTER

Marilyn Chow, R.N.

Traditionally, health services in child-care programs have been limited to physical examinations upon entrance and morning inspections thereafter. Parents are left to their own devices whenever their child becomes ill while at a child-care center. At a time when a child and his family are most vulnerable, they are too often left stranded for adequate health services. When a child becomes ill, his parents typically incur the following problems: (1) the parent must take time off from work; (2) the child must be taken to see a doctor; and (3) the parent must either locate another source of child care while the child is ill or remain at home. Because of these problems and inconveniences, parents frequently neglect their child's illness at a most critical time in his development.

Health services in child-care centers must be concerned with the child's needs for love, attention, and protection. That is, health services should provide for the emotional and social aspects of the child's development as well as for his physical well-being.

Health programs must also integrate the culture of the parents and children in the particular program. Bilingualism should be included as well as the traditional medical culture of the family wherever possible. The health program in a child-care center should therefore consider broad and comprehensive aspects of the child's development, and should involve the children, their families, and the program staff. In addition, the formation of a health advisory board composed of members of the center staff, parents, and health personnel is mandatory.

The purpose of this chapter is two-fold: to present general guidelines in planning a health component for a child-care center and to share the author's personal experiences in the planning and implementing of a health component.

HISTORY

During 1971–72, a health program evolved at the Haight Ashbury Children's Center (HACC) in San Francisco. The HACC has been a federally funded, preschool program for children of low-income, working parents. As with most day-care centers, a health component was required according to federal guidelines, but the mechanism for planning and providing for this health component was not defined.

Consultation was sought from the School of Nursing at the University of California, San Francisco. With the help of the Director of the Pediatric Outpatient Department, pediatric consultation and service was provided every Thursday morning by interns, residents and a pediatric fellow. Consultation was always available in case of emergency. Between September, 1970, and June, 1971, a pediatric fellow was the primary physician designated for the

HACC. In addition, a parent who had recently completed training as a nurse volunteered her services to the HACC on a half-time basis.

This staff performed certain health screening procedures; however, the parent-nurse could not continue her work as health coordinator on a long-term basis. Although the pediatric fellow continued to visit HACC on a service basis, the absence of the parent-nurse emphasized the need for someone who could assume ongoing responsibility for the health component. In July, 1971, a new pediatric fellow arrived to continue the health services at the HACC. In collaboration with the new pediatric fellow, I began to develop my role at the HACC as nurse/health coordinator.

ROLE DEVELOPMENT

At the HACC, I served as health coordinator. I was trained as a pediatric nurse practitioner (PNP), a background which I feel is essential to the role of health coordinator in a child-care program. The PNP concept is relatively new, and particularly so with respect to the area of child-care programs. In the child-care setting, the role of the PNP is flexible in regard to the program's goals and the child's needs. During the past year (1971–72), several models of the PNP in the child-care setting were developed. One involved the use of the PNP in an infant center for teen-age mothers, another was used in a respite center, and two others were developed for preschool day-care settings. In each of these environments, there was a unique development of the health care program and the role of the PNP.

One of my first projects upon arrival at the HACC involved defining my role as a member of the staff. In the beginning, my role as a health professional was rather vague. Because of my interest in the areas of child growth and development, I participated in the staff's curriculum planning sessions as well as the planning for the staff development program. I hoped to develop a total health program which would provide services to the children as well as incorporate a strong educational component for the children, their parents, and the child-care staff. My goal was to work with all three groups in establishing a program that would meet a child's present and future health needs, his parents' interest and involvement with these needs, and the staff's ability to provide the necessary services and information.

For the first three months, my main role was to determine the degree of illness in the children, administer first aid if required, and refer them to a physician. I also determined the health status of the children by reviewing their health records and arranging health screenings when necessary. As I gained the trust of the staff by demonstrating my capabilities in these functions as well as in the classroom, working with the children and the staff in preventive health measures, I assumed greater responsibility and used more initiative in defining my role. The process of developing and implementing a comprehensive health program was one of slow evolution and required constant effort on everyone's part. A major project in this effort was to develop an accurate and useful health record system. This provided a solid basis for functioning in my role with greater confidence and coordinating various activities involving the parent and the staff.

PROCEDURES

Record System

Obviously, a record system which provides an accurate and complete description of the child's health history and his current needs is essential for the operation of a comprehensive health program at a child-care center. The HACC did maintain health records on each child, but they were often incomplete. In reviewing the children's health records at the HACC, it was frustrating to discover inconsistencies in the follow-up of health admission standards. For example, some children admitted into the center's program did not have a recent tuberculin skin test, and only limited attempts were made to insure that a skin test was obtained soon after admission.

Another source of frustration was that frequently a physician was listed in a child's record, but information was not available about the child's health from the physician. Often, the physician listed on the child's record was not known by the parents on a regular basis.

Obtaining immunization records proved to be another frustrating task. While parents were requested to provide such information, usually an immunization record had never been obtained—or it was lost. It was also discovered that most children never had undergone any of the usual health screenings.

Thus it was necessary to organize a health record system to insure that a complete health record was on file, and current for each child in the program. A workcard which served as a quick health index was used. It contained the following information on each child: name, birthdate, date of last physical examination, vision, hearing, speech, dental and developmental screenings, tuberculin test, immunization status, lab tests conducted, and a miscellaneous section in which to note such items as the private physician and other pertinent health information. This card, when completed, provided ready reference to essential information for the preventive health maintenance of a child.

Forms which were included to provide a more complete health record included a health assessment form, a physical exam record, a growth chart, an immunization record sheet, a screening results sheet, and history sheets for progress notes.

In retrospect, I was dissatisfied with the system because it did not include a section listing the child's problems in an organized manner. I later discovered that the Weed problem-oriented approach to recordkeeping (1) was the answer to my problem. Essentially, Dr. Weed's approach advocates the establishment of basic information to be obtained on each person, formulation of a problem list based upon the gathered information, plans for each problem, and follow-up on each problem. This approach provides structure, efficiency, and continuity of care for the children.

In setting up a record-keeping system, I strongly recommend a problem-oriented front sheet for each child's record. Table 9-1 is an example of a front sheet which lists typical problems.

Table 9–1
Problem Sheet

Problem	Date
1. Health Supervision	January 4, 1975
a. Family History (any elements which bear upon child's health, i.e. sickle cell disease)	
b. Child rearing (toilet training, diet, etc.)	
c. Development (growth charts, DDST, etc.)	
d. Immunizations needed	
2. Chronic ear infection	February 2, 1975
3. Iron deficiency anemia	May 4, 1975

Each health record should consist of the following:

1. Front sheet which lists the problems so that anyone can easily determine the scope of the child's problems.
2. Health assessment form to be filled out by the parent(s).
3. Immunization record.
4. Growth and development charts.
5. Results of any health screenings and laboratory tests.
6. Progress notes.

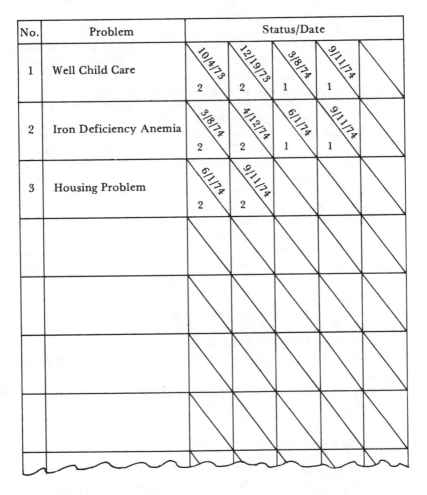

Fig. 9–1
Sample
Problem Oriented Medical Records

Children's Center

Name:_____

Address:_____

_____

Chart No.:_____

Problem List

No.	Problem	Status/Date				
1	Well Child Care	10/4/73 — 2	12/19/73 — 2	3/8/74 — 1	9/11/74 — 1	
2	Iron Deficiency Anemia	3/8/74 — 2	4/12/74 — 2	6/1/74 — 1	9/11/74 — 1	
3	Housing Problem	6/1/74 — 2	9/11/74 — 2			

Status Categories: 0 = Inactive; 1 = Active Stable; 2 = Active Unstable

Fig. 9–2

Children's Center

Patient Name_____ Date_____

Problem #1   Well Child Care

Subjective

a. Family: main family concern discussed in Problem #3. Warm, stimulating environment in home. Mrs. D. has been spending more time with kids on weekend and is getting out herself more than before.

b. Child Rearing: Mrs. D. no longer having significant concerns. Has been using discipline more judiciously and consistently. Finds that sitting down and reasoning with Jonathan pays off.

c. Nutrition: well balanced diet with adequate sources of protein. Mrs. D. has tried cutting down on sweets and has iron-containing foods for whole family.

d. Development: see Denver Developmental Screening Test in chart. Normal language development for age. Teachers report Jonathan socializing well at center—much less crying than last year.

e. Child Safety: Mrs. D. has not been able to get harness auto restraint for Jonathan.

f. Immunizations: complete for age. Will need boosters before starting kindergarten.

g. Dental: brushing teeth fairly regularly—older sister is responsible. Mrs. D. has made appointment for Jonathan to see dentist on 9/28—District Health Center.

Objective

Height: Weight: see growth chart.
Physical Exam: normal
Vision Screen (Titmus): R 20/30 — L 20/30
Hearing Screen: not done because child not cooperative—very tired.

Assessment

Problem is stable. I feel that things are going relatively well in family at present.

Plan

a. Repeat hearing screen 6 months.
b. DPT booster in 6–8 months.
c. Encourage Mrs. D. to follow through on dental appointment.
d. Encourage to get auto harness restraint and to keep kids in back seat until then.

Fig. 9–3
Problem Progress Form

Children's Center

Patients Name_____ Date_____

Problem #2   Iron Deficiency Anemia

Subjective
   Mrs. D. is regularly giving iron-containing foods—red meats, eggs.
   beans, raisins, etc.

Objective
   repeat hemoglobin on 9/1 was 11.0 GM %; retic. count was 3%.

Assessment
   problem is active stable.

Plan
   R: continue to include iron-containing foods in diet.
   Patient Education: have reassured Mrs. D. that the problem is not a
   serious one and that there is significant improvement.
   Follow up: repeat Hb in 6 months.

Problem #3   Housing Problem

Subjective
   The apartment situation is the same. Space is totally inadequate for
   5 persons. Mrs. D. has not found time to look for a larger place
   but plans to look this weekend. She's pessimistic that she'll find
   something that she can afford.

Objective
   Made home visit last week. Apartment is well kept despite lack of
   space. No medicines or household toxins are kept where children
   can get them.

Assessment
   Problem unstable

Plan
   a.   have arranged for Mrs. D. to speak with center social worker
        about possible housing resources.
        Follow-up: will contact Mrs. D. by phone in 2 weeks.

The information in the records is confidential. The records themselves are perhaps best kept in a file cabinet located in the appropriate office.

Health Assessment

A comprehensive health assessment program involves an ongoing process of assessing each child's health, taking the required action, and recording it in the child's health record. The value of the child's physical examination upon entrance to the program declines rather quickly with the passage of time, particularly if the program does not insure periodic and systematic review of the child's health. To achieve such a program, teachers can be trained to observe and record observations based upon the child's classroom behavior, and health personnel, such as pediatric nurse practitioners, should be available to the program on a permanent basis.

A total health assessment for a child encompasses a health history, which includes a discussion with the parent(s) and a record of immunizations; a physical and developmental assessment; and health screening procedures, which include laboratory tests.

The discussion with the parents should focus on questions they have concerning their child, such as parent-child relationships, problems in child-rearing, and any other concerns identified by the parents.

Immunizations received should be recorded on the immunization record; a tuberculin skin test completed within the last 12 months should also be documented. The health screening program should include vision, hearing, speech, and dental screening as well as urine and blood samples for laboratory testing. Additional laboratory testing should be conducted for sickle cell anemia and lead poisoning depending upon the population involved.

It is essential that an effective referral system be a part

of a screening program to insure follow-up of any abnormalities detected. This raises the issue of a formal linkage between the health services of a child-care program and a structured health care delivery system, for example, outpatient clinics.

Useful aids in determining health standards and the factors of the health assessment include the following two publications by the American Academy of Pediatrics: *Standards for Day Care Centers for Infants and Children under Three Years of Age* and *Standards for Child Health Care* (2, 3).

Community Resources

Immunizations were arranged through the Maternal Child Health Division of the San Francisco Public Health Department. Consent forms are usually required of parents. If a child-care center nurse is unavailable to do the immunizations, the local health department should be consulted. An immunization clinic can probably be arranged to be conducted at the center.

The Easter Seal Society was contacted for speech and hearing screenings. While it was necessary to make the arrangements in advance, an unusually thorough service was provided. The society brought their mobile van to the center and provided a complete test report to the center as well as an explanatory report to each parent. Arrangements were made for a speech pathologist to come to a staff meeting to discuss the results of the screening.

In addition, the speech pathologist presented useful exercises which the children could perform to correct certain problems and provided a list of references for further information.

A local school of dentistry was contacted to make arrangements for a dental screening at the school. One of the dentists on the faculty donated three mornings to screen all the children.

Through the University of California College of Optometry, two students interested in conducting a community project on preschool children volunteered their services. Using a method specifically designed for preschoolers, they obtained comprehensive and valuable information that the standard acuity test might have missed.

Each of the above three screenings was conducted over a period of two to two-and-a-half days. Children who were absent on the screening days occasionally encountered difficulty in rescheduling their screenings, and, unfortunately, some were not screened. Immunizations were also begun during this same period.

Other community resources also can help. For example, the Society for the Prevention of Blindness will provide the necessary materials for a vision screening as well as additional material to which the staff and parents can refer. The Dairy Council of America has attractive posters available on dental hygiene, as well as pamphlets on nutrition.

Obtaining laboratory services may be a difficult task. It will probably depend upon the community's resources. If the use of private doctors is common, then as an entrance requirement it would be useful for the children to obtain a urinalysis and hematocrit through their private physicians. If not, local clinics might be willing to make an arrangement at a minimal cost.

It is also important to locate a source for emergency care. For example, if the child becomes seriously ill or has an accident, where will he be taken that is both accessible to the center and agreeable to the parent? Working out an arrangement at a nearby health facility is advisable.

Following the screening program, any abnormalities detected should be referred to the proper health facility or organization for treatment or correction. This need emphasizes the importance of one person assuming the responsibility for pursuing necessary follow-ups on the screenings.

Policies

A common problem facing a child-care center staff is the formulation of an appropriate policy concerning children who become ill during the day. Sending a sick child home does not adequately resolve the problem, since all too often the parent is unable to leave work, and thus the child is left unattended at home.

A series of respiratory studies conducted in Chapel Hill, North Carolina (4) demonstrated that children in a child-care center did not incur a greater risk for illness than children in a home care setting. For one thing, by the time a child exhibits signs and symptoms of most illnesses, the other children already have been exposed to his illness. Moreover, symptoms indicating more serious illness are detected earlier when the child is being watched by trained health observers. It must be emphasized, however, that a policy of accepting sick children demands adequate staff training and health care.

In regard to the formulation of this policy, there are relatively few instances when isolation is necessary. Obviously, if such cases as chickenpox occur, the child should remain at home. However, in most cases, isolation is just a means of providing the ill child with a quiet refuge from the hustle and bustle of program activities. Thus, a warm, quiet room furnished with a comfortable cot, pillow, and blankets could be set aside where the teacher can frequently check on the child's comfort.

A policy which automatically requires a morning inspection should be reevaluated. In my opinion, morning inspections are of little value, particularly if the person conducting it is not familiar with the individual children. For the most part, the child's teacher is the best observer since the teacher knows the general behavior and activity of the child. In working at the HACC, I found that the teachers would often indicate to me when a child was not feeling well. Typical comments were: "She is not as active as usual," or "He usually eats a great deal, but today he wouldn't eat his breakfast." These indications of probable illness are more useful than a morning inspection of the throat.

Another policy consideration is whether to use such developmental tests as the Denver Developmental Screening Test. The DDST (5) tests the child in the following four areas: personal-social development, fine motor-adaptive development, language development and gross motor development. In addition to screening for any developmental lags, the value in utilizing a screening tool such as the DDST is to help parents become more aware of developmental patterns and to use the tool as a basis for discussion with them. A more refined screening tool, such as the Thorpe Test (6), is valuable in identifying those preschool children with developmental lags who should be referred for a more in-depth evaluation. Developmental screening performed on a routine, periodic basis is of particular value in identifying those children who potentially may have a learning or developmental problem.

Developmental testing, however, has generated considerable controversy because it often contains built-in cultural biases. Therefore, in determining whether developmental testing should be used, the following factors should be considered: possible parental objections; the purpose of the testing and the use of the results; and whether to test routinely or only on referral from the teacher. In any case, a developmental appraisal should be done on each child.

COMPONENTS

Children

The program for the children was the most exciting, primarily because I was given the freedom to plan activities, and the children were enthusiastic and receptive. The goal in working with the children was to increase their familiarity with foods, common medical procedures, and health concepts. For this component I fortunately had the assistance of two nursing students at the University of California, San Francisco. Together we sought the support and cooperation of the staff by soliciting their advice in staff meetings. Through these meetings, field trips were arranged and class projects were developed for the children. Nutrition was a major concern. We exposed the children to foods that many had not eaten at home. They were encouraged to feel, smell, and taste both raw and cooked foods. These classes not only helped to integrate the health unit with the teaching staff, but established our role in the classroom setting with the children.

Many opportunities exist in the everyday program of the center to incorporate health concepts with other activities. Mealtimes, for example, provide a multitude of opportunities to teach children about healthful eating. It is also an excellent time to help children develop their language skills as well as their senses of taste and touch; by discussing where foods come from at the breakfast table, the child's world expands. This means of integrating information received from other sources enables the child to gain greater understanding. Other daily healthful activities should become part of the routine activities of the program. For example, before each meal, the children should wash their hands. If the children are already doing this at home, this activity is reinforced and demonstrates to the child that being in a child-care program is an extension of his home life rather than a separate, foreign experience. On the other hand, if such activities are not being done at home, then this experience in the child-care center may be the genesis of subsequent behavior in the home.

The health program also involved coordinating dramatic play and medical procedures, such as finger pricking for blood testing. The concept of dramatic play in helping the child cope with intrusive procedures is crucial to the health program. Prior to a mass sickle cell and hematocrit screening, we felt that it would be important to have the children play with the equipment to be used for the screening, such as capillary tubes, gauze squares, and red colored water (substitute for blood). For two consecutive sessions, the children were given the opportunity to play doctor and nurse. As part of the play, the children "pricked" each other on the finger, then drew up the colored water into the capillary tube. We explained what was happening step by step and we had the children play similarly. In the second session of play, we did an actual finger prick on a teacher, who then told the children how she felt about having her finger pricked. Some of the children became frightened, but others watched carefully. These sessions were helpful in that they gave the children an opportunity to work through their negative feelings about having their finger pricked. When the actual finger pricking occurred, the children were most cooperative, confirming my belief in the value of dramatic play as an effective tool in a health program.

The health program for the children used knowledge of growth and development to enhance the children's understanding of sound nutrition. Furthermore, their coping mechanisms for dealing with intrusive procedures, such as anemia testing, were allowed to develop naturally, thus minimizing negative reactive feelings.

Parents

Parents have a multitude of concerns about their own personal lives and their jobs as well as their children. How realistic is it to expect that parents devote much time to the child-care center? How much time can the center demand from them without being judgmental of their role as parents? What are realistic expectations of them concerning their interest and involvement with the child-care center? These questions are not easy to answer, and yet they must be considered in the planning of any child-care program.

Prior to my arrival, parents were not an integral part of the program because of the difficulty working parents experienced in devoting time to the center's activities. Usually, they dropped their children off before work and quickly picked them up after work to hurry home to household chores. This left little time to discuss concerns and problems regarding their children.

Planning for the involvement of parents in their children's child-care program proved very difficult. At the HACC, monthly parent meetings were arranged by the social worker and a parent committee. Before each meeting, the parent committee personally notified each parent of the upcoming meeting and the importance of their attendance. A parent newsletter, which included a child health column, was published every two months to supplement the parent meetings as well as encourage attendance at them. Yet all too often only a few parents would attend these meetings, a continuing disappointment to all those who arranged and conducted them.

One of the more successful methods of stimulating attendance, however, consisted of relieving parents of their pressing concern of preparing dinner after their work day. Occasional potluck suppers at the center provided a relaxed atmosphere for parents and staff alike, as well as a natural setting for discussions of nutrition, a vital component of the center's health care program. Question and answer sessions dealing with preschoolers' eating habits, allergies, and vitamins always generated considerable interest. If parents have more detailed questions, a nutritionist could be invited to speak to them. The local school department and state child-care center programs may be of help in providing such a person.

If the staff is limited in time, an alternative idea for parent involvement is to use weekend workshops. These can be organized around specific topics such as nutrition, "when to call your doctor," or just a general discussion about topics on which parents want more information. At another child-care center in San Francisco, this idea is being used very successfully. The parents are very verbal and willing to ask questions. Ideally, parents should be included in the planning committee, so that their input will be available from the beginning of such a program.

The question arises as to the kind of handout information parents should be given on the care of common illness, such as fevers, vomiting, and diarrhea. I prefer to distribute handouts to parents only after health classes or discussions to insure that the handouts will be read and to provide for the opportunity to clarify any misunderstandings immediately.

When planning for parent involvement, the staff must consider the particular population with which it is working. In view of time and interests, what is feasible in terms of parent involvement and education? Perhaps a parent committee can be organized to plan together with the staff meetings and educational sessions. Communication is essential to the functioning of a viable health program.

Staff

The program for the staff consisted of working with each staff member on an individual basis as the need arose, rather than in regularly scheduled classes. This was largely due to time constraints.

At staff meetings, however, interesting films concerning nutrition and child growth and development were shown occasionally. These were obtained from the Dairy Council of America and Modern Talking Pictures.

The Red Cross conducted first-aid classes for both staff and parents and the center provided child care for the parents during these classes.

Parent education programs can also be a valuable means of educating staff members. At one nursery school, a parent meeting featuring a nutritionist was also attended by the staff. They were able to get answers to such questions as what types of snacks should be served, how to make meals more interesting, and how to plan for meals.

SUMMARY AND CONCLUSIONS

The provision of health services as an integral part of a child-care program is vital. It is not an easy task. It involves communication, coordination, and organization among the child-care center, the health care delivery system, and the family. If we are committed to the health of our children, we must seek and provide for their optimal growth and development.

PROGRAM RESOURCES

A. Places to visit
　　1.　Hospital tours—children's hospitals and clinics
　　2.　Dental tour—public health department
　　3.　Ambulance—city ambulance
　　4.　Private doctor's office
　　5.　Prepare a health corner with exhibits, props, nutrition displays, facilities for dramatic presentations.
B. Screenings and referrals
　　1.　Dental—dental schools, public health department, local clinics

2. Vision—Optometry school, mobile clinics, fraternal organizations such as Lion's Club, Police Athletic League
3. Hearing and speech—Easter Seal Society
4. Sickle cell anemia and hematocrit—medical school laboratories.

C. Record System

To provide accurate account of child's health status, utilizing tools quickly to assess the health needs of each child.

1. Pediatric history questionnaire
2. Immunization record
3. Progress note sheets
4. Miscellaneous reports such as dental and vision reports
5. Health program control sheets as developed by the Office of Child Development
6. Card index file for quick checks.

D. Other Resources

1. Dairy Council of America
2. American Red Cross
3. School of Optometry
4. Department of Public Health
5. American Heart Association
6. Easter Seal Society
7. Regional educational laboratory and/or local school system
8. Local ambulance service
9. Colleges and Universities
10. Modern Talking Pictures
11. Local school of nursing
12. Nutritionists
13. Public Health Nurses
14. Social Workers
15. Physicians
16. Dentists
17. Occupational Therapists

18. Early Childhood teachers
19. Office of Child Development, U.S. Department of Health, Education, and Welfare, Washington, D.C.
20. Day Care and Child Development Council
21. National Association for the Education of Young Children
22. Association for Childhood Education International. (See also resources in Appendix.)

NOTES

1. Weed, Laurence. *Medical Records, Medical Education, and Patient Care.* Cleveland: Case Western Reserve University, 1969.
2. American Academy of Pediatrics. *Standards for Day Care Centers for Infants and Children Under Three Years of Age.* Evanston, Ill., 1971.
3. _____. *Standards for Child Health Care.* Evanston, Ill., 1972.
4. Peters, Ann DeHuff. "Health Support in Day Care." In *Resources for Decision*, ed. by Edith Grotberg. Washington: Office of Economic Opportunity, 1972. Pp. 315–339.
5. Frankenburg, W. and Dodds, J. "The Denver Developmental Screening Test," *Jour. of Pediatrics* 71 (Aug., 1967). Pp. 181–191.
6. Thorpe, Helene. *Developmental Appraisal of the Preschool Child, Ages 4–6.* Davis, Calif.: University of California, in press.

BIBLIOGRAPHY

American Red Cross. *First Aid Textbook.* Garden City, N.Y.: Doubleday.
Brandon, Brumsic, Jr. *All About Healthy Me.* Washington: Department of Health, Education, and Welfare and Office of Child Development, Project Head Start.
Dittman, Laura. *Children in Day Care with Focus on Health.* Children's Bureau Publication, 1967.
Harrison, Dorothy. *Healthy, That's Me: A Health Education Curriculum Guide for Head Start.* Washington: Department of Health, Education, and Welfare and Office of Child Development.

Lictenberg, Philip and Norton, Dolores. *Cognitive and Mental Development in the First Five Years of Life.* Washington: National Institute of Mental Health, 1970.

North, A. Frederick, Jr. *Health Services: A Guide for Project Directors and Health Personnel.* Series on Day Care, no. 6. Washington: Office of Child Development, 1972.

Problem-Oriented Medical Records

Aradine, C. and Guthneck, M. "The Problem Oriented Record in a Family Health Service." *American Journal of Nursing* 1.74: 1108–1112.

Bjorn, John C. and Cross, Harold, *The Problem Oriented Practice of Private Medicine.* New York: McGraw-Hill, 1971.

Bonkowsky, Marilyn. "Adapting the POMR to Community Child Health Care." *Nursing Outlook* 20 (Aug. 1972): 515–518.

Hackson, Robert and Morton, Jean, eds. *Evaluation of Social Work Service in Community Health and Medical Care Programs* (Part III POMR). Berkeley, California: Institute for Public Health Social Workers, University of California.

Weed, Laurence L. *Medical Records, Medical Education, and Patient Care.* Cleveland, Ohio: Press of Case Western Reserve University, 1969.

Chapter Ten

A PSYCHOLOGIST'S VIEW OF COMPREHENSIVE SERVICES

Dorothy Nash Shack

Child psychologists take as a given that all children do not conform to predetermined and standardized models of development and learning. Yet however minimally individual growth patterns might reveal deviance from the expected or "normative" measurements of development, such differential variations traditionally have been viewed as the child's inherent incapacity to learn. The child psychologist, therefore, has been thrust into the role of establishing the fact of differential development, not detecting the cause of it. By confirming the child as the "problem," as such analysis has shown in other areas of life, the child psychologist often has so distorted reality that solutions to impediments to learning often have been sought in the wrong areas. Psychologists now are moving more decisively in the direction of providing early detection and prescriptive services to young children that can undergird strong child-oriented educational programs. Of prime importance in this effort is the real acceptance by society of the individualistic patterns of human development that underlie much of what is labeled failure, inadequacy, or incapacity.

Day-care centers provide the earliest opportunity for a social institution, such as the school system, to deliver to children detection and prescriptive services that can promote their optimal development, thereby minimizing the difficulties children face in learning to cope with the demands that society ultimately places upon every individual. From age two, and often earlier, children are enrolled in day-care centers, spending up to eight or more hours per day in a peer-group setting with several adults. In many centers today this population of children is heterogeneous with respect to race, racial integration now being an accepted way of life in many school systems. Often, sincere efforts have been made to enable teachers to achieve some knowledge of the variety of cultural patterns they will have to understand and respond to when teaching in multiethnic settings. This also has been true for parent groups and parent-teacher groups. No one should expect that in only a few years all the misunderstandings, conflicts, and preferential treatment due to race would end. But teachers are becoming increasingly sensitive to their own biases and therefore are moving well in the direction of interrelating equitably with all of their pupils.

In the day-care centers, teachers are assigned the role of providing multiple services for youngsters in both the simplest and most complex facets of child care. The increase in day-care facilities has created a need to employ new teachers, many of whom come to the job minimally prepared and who usually are registered in at least one in-service and/or college course per semester. Programs for on-the-site assistance are a great help to these inexperienced teachers as they struggle to cope with often overwhelming daily problems.

This article will address itself to basic issues underlying psychological services to day-care centers to insure the optimal development of each child; a program of professional services that moves decisively away from the conventional notion of custodial care.

The Early Growth Center in Berkeley, California, provides an educational program based on detection and prescriptive services to young children. The basic method for achieving this program's goal is through teacher training and parent education. The primary goal of the program is the integration of children with handicapping conditions into the "normal" preschool population and to assure the success of this process by providing individual program plans for each child in the school.

In-service education is provided for teachers. A model preschool has been developed in a Berkeley Unified School District day-care center where a one-month internship is available to teachers in the Berkeley Early Childhood Education Department. Ideally, teacher-interns should return to their schools more committed to the integration of children with handicapping conditions into the preschool, more capable of providing for the needs of children of all descriptions, and with a more positive approach to providing for the needs of children of various ethnic origins. The psychological services provided to the program can be divided into several areas, the following of which will be discussed more fully:

1. Support and guidance in rethinking the framework of the school

2. Provision for the development of appropriate behavior in children and positive movement toward maturity

3. Establishing positive, ongoing relationships between parents and teachers.

Rethinking the Framework of the School

The responsibility for defining the relationship between the child and adult caretaker rests with the adult. It is, indeed, an art form to interact so that child and adult both enjoy the relationship, that they experience minimum strain because of it, and that the emotional ties between child and adult are appropriate and strong.

Children sense or understand their dependency on adults and are confused, and often frightened, when permitted too easily to dispense with adult control. In the latter case they then must make decisions regarding their own role in the home, school, and society generally, or at the extremes they withdraw from growing up. To circumvent this premature assumption of independent behavior, adults in charge of young children must make decisions about who controls the relationship, what the rules are that govern that relationship, and how the rules are to be enforced. Acceptance of this posture for adult-child interrelations leads to the issue of communication between the parties. From the first day of school, teachers should communicate to all children, in a clear and positive manner, what can be provided by the school for the children and what is expected of the children; in other words, the ground rules should be made clear.

In order to communicate rules of procedure to children, teachers themselves must be in agreement on all major areas of concern in the school. They must be willing, as cooperating adults, to be flexible in terms of their own attitudes and beliefs about children and child care. Teachers should present to children a united front: the reality of a group of adults, differing from the parents, yet concerned with the children's welfare; prepared to provide for, care for, and protect the children during school hours; dependable, flexible, and enjoyable. Early communication to children of this reality by all the adults involved establishes the basis on which good teacher-child interrelations can be built.

The psychologist in the day-care center can serve as a force to help teachers move toward this position. The procedure for achieving unity among teachers within the same school is complex, but very rewarding, for early in their efforts to restructure their thinking and behavior toward children the rewards are seen in the decrease in disruptive behavior and the greater tendency of children to cooperate with a plan both reasonable and productive. Regular staff conferences in which teachers discuss specific daily issues and problems have comprised the traditional means by which the staff evolves attitudes and behavior to give consistency and sensitivity to a center's program. Proper direction given to these meetings by the psychologist can aid the staff in developing a frame of reference within which to classify and deal with "problems" on other than a piecemeal basis. In this way the staff can anticipate the development of certain situations and act before they evolve into real problems.

The modeling of responses to specific behaviors demonstrated by children leads to the development of a larger repertoire of interactions around a single type of behavior which teachers have heretofore felt incapable of dealing with satisfactorily. Teachers are relieved of the necessity to respond "correctly."

The organization of the school day also can be made flexible, so that in influencing the flow of children into available spaces, stress situations not only can be reduced but also more easily be relieved.

Not in answering questions from a lofty professional perch, but in asking about the mundane aspects of the total program, the psychologist functions to get teachers to rethink the whys of habitual organization. They find "we've always done it that way" an increasingly unsatisfactory answer.

Provision for the Development of Appropriate Behavior in Children and Positive Movement toward Maturity

What is appropriate behavior? First it is dependent on age; second, on what society defines as acceptable behavior for a particular situation at a given age. How children do, in fact, behave depends on the training received up to the time in question. Young children often find themselves confused by the fact that behavior which formerly won praise may now result in their being scolded. This is, however, a fact of life that children must learn. Schools are responsible for reinforcing normative behaviors as well as for dealing with the confusion and resistance children exhibit toward what may well be real inconsistencies in societal expectations.

Encouragement of appropriate behavior requires the utilization by adults of specific techniques to foster maturity of children, thereby promoting the gradual assumption of responsibility for personal conduct. The particular techniques to be used in this effort could be based on one of the following positions:

1. Adults should take full responsibility in assuming the role of authority figure. That means making all decisions concerning the health, welfare, and education of youngsters in their charge. Such a position of authority belongs to the adult whose task is to guide youngsters through their formative years; adult experiences give preparation for making certain judgments about a child's welfare in a world too complex for the child to cope with.

2. For children, the world, even the home environment, is a large, unwieldy, and indecipherable place. Trying to find their way in so complex a world is a formidable task for children. Unable to

grasp the meaning of the signposts along the way, children can become increasingly confused and upset for reasons they cannot comprehend and over which they have no control. Children need firm guidance in order to learn to live in their own environment without undue stress and strain.

3. The conclusion to be drawn from (1) and (2) above is that adults must structure the child's learning situations so as to permit the gradual perception of the complexity of life and ways of handling it without undue anxiety, difficulty, or permanent damage to self or others.

In the effort to effect this kind of adjustment to life, from infancy on there must be limits placed on both the geographical and psychological area to which children have access. These limits must be flexible; they should be moved in varying directions and at varying rates. The purpose of the limits is to provide a control system within which the child can safely and securely begin the exploration of his world. What I here call the "limit-control system" might be presented schematically as follows in Fig. 10–1.

The detached wall structure, each section of which represents a major area of development, implies a discrete separation of basic growth areas. This is a false separation —the rounded corners are imposed to indicate the interrelationship among the developmental areas. The broken circle surrounding the whole represents the child's niche in the total scheme of ever-changing life forces. Changes in the relationship between the child and his environment can move either positively—the limits on the accessibility to the child of the world at large are steadily decreased; or negatively—the limits on the accessibility to the child of the world at large are steadily increased so that growth is inhib-

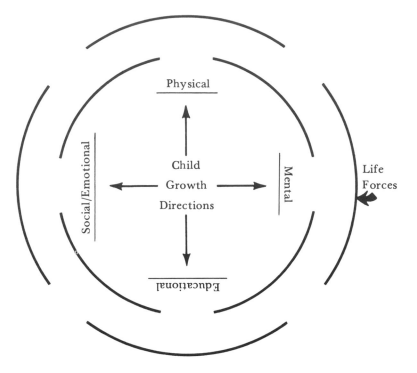

Fig. 10–1
Limit-Control System

ited and the individual severely handicapped in the ability to make a positive adjustment to life.

So also the movement occasioned by the change in the child-environment relationship can be instigated from one of several sources:

1. The child, in the normal growth process, is hampered by the limits and by virtue of the need for greater latitude; he himself initiates the outward movement of the limits.

2. The adult, feeling the need to stimulate the child's growth process, extends the limits as a motivational step.

3. The vicissitudes of life also bring about change in
    the limit-control system. A traumatic event occurs
    to the child. This may have the effect of an explo-
    sion, blowing out the walls of the limit-control sys-
    tem. The walls may return to the original position,
    may move in closer, or move in, but not so far as
    to return to their original position. Finally, walls
    may remain drastically expanded, which puts the
    child well out into the environment prior to learn-
    ing the methods for succeeding or surviving in that
    environment. In this third category, contrary to
    the first two presented, the factor precipitating the
    change in the system is something outside of the
    realm of control of the adults in the child's life,
    such as death, illness, injury, fire, flood, or trau-
    matic experiences with unknown persons or
    things.

The purpose of the limit-control system is to give children some resistive force against which to push in the
growth process, and provide them with some information
about the world which can be used to begin to make associations, judgments, and decisions with some degree of success. In short, children are thus provided with a framework
for dealing with day-to-day life situations. They then are
able to meet their own needs and be comfortable in the
ways elected to do so; able to obtain the greatest good from
the expenditure of energy, establish positive human relationships, respect individuals and their rights and, in turn,
be respected as individuals.

The utilization of such a system with many children,
or, for example, in a school setting, is dependent upon the
amount of information teachers have about each child and
how this information is shared with other teachers, parents,
health consultants, and guidance workers in establishing a
descriptive profile on the child that indicates how that child
is functioning in all areas and that leads to the development

of a specific individual program plan for promoting developmental progress. Teachers following such organization soon will come to realize how easily they and parents can together look at most children, see their strengths and areas of inappropriate response or behavior, and design both home and school plans to move the child toward fuller and higher quality development. In an effort to develop an organization aimed at achieving the above goals, the following general process has been instituted between the staffs of the Early Growth Center and the Martin Luther King, Jr., Children's Center:

1. General observation of all of the children begins the first week of the fall school term. Children having difficulties in the following areas are noted: following rules, peer or adult interrelationships, utilization of the curriculum activities successfully, and aspects of the children's total being that may make them uncomfortable or less successful than their peers. Notation is made also of those children who exemplify the "average"—that is, who are getting along well in all areas of development. Throughout the year these latter children are subject to reevaluation to insure their continued positive growth. Children who have specific difficulties that may be due to lack of preparatory experiences, general immaturity and/or "difficult behavior," or some physical handicapping condition are subjects for detailed plans involving parents, teachers, and all necessary support services. These plans cover the range of school activities and possibly home activities, and are updated regularly.

2. Administration by teachers of the preschool inventory that is required for all children by the school district.

3. General pediatric examination of each new child including screening with the Denver Developmental Scale. The physical examination is followed up with appropriate referral for hearing, vision, physical therapy screening, or other indicated services.

4. Developing acquaintance with parents. Learning something of their life-style and parent-child interaction pattern.

By mid-September, a great deal of knowledge has been gained about the 35 children who attend the Martin Luther King, Jr., Children's Center, some of whom are new to the school, some who have entered during the summer months, and some who are enrolled for a second or third year in the school. Based on information shared in staff conferences, further next steps occur.

Individual children, specifically including those who are, as part of their individual program plans, continuing in preschool rather than entering kindergarten, are selected for closer observation at specific times throughout the year so that development of physical and mental maturity, the utilization of the preschool learning centers, and the appropriateness of behavior can be noted.

Individual children considered "at risk" (children whose experiences have not prepared them for success in the school system) are selected for small group (two or three) or individual work in the curriculum area.

Parent/psychologist conferences are held to elicit additional information about the child to determine how the parent views the child's general development and adjustment and to discover what are the parental expectations for the child. Where indicated or requested, suggestions for home activities and behavior-development techniques are discussed, as well as personal problems parents may be facing.

Staff discussions focus on these specific children to determine how the school can best promote optimal development in all areas.

A program plan is designed for all children, including those who appear to be developing within "normal" limits. Part of this process is regularly scheduled review, at which time the individual plans are updated to suit the rate at which the acquisition of skills, knowledge, and socialization is taking place.

By January the staff has become increasingly concerned with the progress being made by four-year-olds. The total development and adjustment of these children is regarded in anticipation of entry into kindergarten the following fall.

By March a preliminary decision is reached for those children who, considering their current adaptation to preschool, most likely will experience undue stress and/or failure in kindergarten. Conferences are held with parents of these children, detailed discussion of the child's progress and areas of learning or behavioral difficulty is carried out, followed by a strong recommendation that the child spend an additional year at the preschool level. The staff regards these decisions flexibly, realizing that by September such a child may indeed be ready to enter kindergarten—or, if not, perhaps to begin to use successfully some kindergarten-based materials within the preschool program.

The process in the school is, then, one of carefully providing for children's growth in all areas, making certain adjustments for specific children, reviewing and updating of individual program plans, and counseling with parents in regard to specific current needs of the child, and the role of the home in promoting development.

Throughout this evaluation procedure, the psychologist strives to gain teacher and parent acceptance of the fact that they indeed can interpret the children's individual developmental needs and can suggest positive, preventive measures. In so doing, the aura of mystique surrounding the psychologist's everyday activities should be removed, thereby instilling confidence in parents and child care workers that they, too, can develop the skill to assess and provide for the needs of most children.

Establishing Positive, Ongoing Relationships between Parents and Teachers

One component of the Early Growth Center program is a parent discussion group which is intended to provide a forum where staff can effectively provide general information to parents on some basic aspects of child development and parenting, and parents can learn more about the total goals of the school and how the many parts of the program form a meaningful whole. Parents are able quite meaningfully to share with their peers the joys, trials, and tribulations of having young children and come to respect the many life-styles seen within the group. From this perspective they are better prepared to make judgments on their own life-style in terms of its positive and negative effects on child-rearing and in relation to their hopes and aspirations for their children. This is of the utmost importance. Many parents can articulate the type of adult they would like their child to become. However, often they have not thought through the effects on that ideal made by their current child-rearing practices.

In addition to these aspects of the parent program, there is an impetus to increase the benefits of parent-teacher interaction for children, parents, and teachers. The difficulties experienced by both parents and teachers in their interactions around the welfare of children appear to have several enmeshed strains:

1. The need of the parents for their children to succeed in relation to their peers, which, among other rewards, gives social approval to their way of child rearing.

2. The need of the teacher to have each pupil progress according to the preset expectations, for this proves the presence and quality of the teaching.

3. The full knowledge that teacher and parent have of their own needs, and the vulnerability of each to peer sanctions when the child has not met the standards, result in negative undercurrents in the parent-teacher relationship.

4. Teachers often experience difficulty discussing children's areas of need with the parents either because of the underlying belief that these "needs" are really gross inadequacies of the children, or that the parent will see the teacher-effort to unravel the problem as a statement of the inadequacies of the child and the parent, for which someone must be held accountable.

5. Parents often experience difficulty in discussing children's areas of needs because such discussion may to them constitute an admission that the child is not in fact progressing according to expectations.

The lead question then becomes: Has the child failed due to inadequate parenting or inadequate teaching? Thus begins a verbal and behavioral sparring match between parent and teacher that can forever cloud the real issue of the needs of the child. Too infrequently are these needs dealt with positively in a forthright manner, and so the cycle repeats itself at an even faster rate as minor issues become complex problems.

To promote parent-teacher understanding and cooperation, the positive approach in all discussions of child-rearing, development, and behavior is of major importance. Staff discussions of children, and indeed all analyses of individual children and their needs, are conducted in terms of how to utilize the strongest areas of development and strongest needs to promote movement in less well-developed areas. Underscoring for parents the fact of the youth of the small pupils and the diversity of their backgrounds helps parents put into perspective any areas of difficulty a three- or four-year-old may be experiencing. Deliberately using descriptive terms that have a positive social value to replace those with negative connotations cannot be underestimated for its value in the reduction of stress.

Thus, a four-year-old boy who came to the Martin Luther King, Jr., school from a setting where he was indeed labeled "extremely aggressive," was seen in the new setting as a very exuberant and strong child. To have the joy and vitality that he showed seen in a negative sense had certainly been destructive to any forward-moving plans. While it is necessary to help such a child bring his vitality into a more harmonious relationship with his total environment, the positive acceptance of him in the King school was a help to all in the planning and living with aspects of his months in the school. The boy himself was at first very confused that teachers were not constantly angry with him, and he queried them about their feelings toward him.

Parents are more receptive to interpretations and suggestions when teachers present them in a positive frame of mind. As they see their children as individuals with individual patterns of behavior, parents are less inhibited in analyzing their own child-rearing techniques.

In a sense, teachers must convince parents of their determination to have children progress to an optimum degree and that all techniques will be utilized to insure this development. At the same time, parents are asking for help in getting their children moving toward optimal development. Thus the psychologist might well profit by employing the techniques listed above, as well as others, in order to encourage parents and teachers to work cooperatively toward achieving their common goal.

What I have tried to show in this discussion is an overview of a preschool program aimed at the individual needs of children. The psychological support system is infrequently labeled as such because the essential job to be done is one of pointing towards directions for movement, opening new possibilities for consideration of behavior, accepting behavior as the individual's best way of coping with the environment, and, through teaching, increasing the options children have for coping with their environment. Special skills of the psychologist for analyzing behavior, working toward the modification of behavior, interviewing, counseling, and testing are all intricately woven into the design of the overall program.

Chapter Eleven

THE ROLE OF THE SOCIAL WORKER IN CHILD CARE

David Brown

I feel at home here. I can wander anywhere I please. . . .
There's always another chair available to pull up to the lunch
table, and my help is usually needed. And my child sees me
eating with her. . . . When I want to relax, I can have a cup
of tea and sit on the sofa in the lounge. Usually there is
something to read, even school news, and people to talk to.
. . . It's a chance to get out of the house, and the everyday
routine. And the kids know me. They come up to me to show
their drawings, or touch their bruises, or just sit on my lap.
. . . I like to see what the children are learning.

This is a composite of remarks from parents in one of the
child-care centers I visit. The center truly represents a
home away from home, where some relief from the stress-
ful world is possible. The ideal center should provide a
supportive, caring, and stable atmosphere where mothers
or fathers, boy friends, grandmothers, or aunts can watch
their children learning and playing and where they can

observe teachers using a variety of techniques in relating to children in constructive ways. A child-care center can engender a feeling of neighborhood and can stimulate a healthy cultural diversity. It can provide a sense of purpose amid the loneliness of city life, the isolation from family and relatives, and the alienation from a confusing and complex society. With the full association of parents and families as equal partners in the education of their children, the child-care center director and staff can instill the values thought to be most vital in the rearing of children of this generation.

As a social worker, I would expect the following characteristics in those who provide counseling, social services, or consultation to children, parents, teachers, and administrators:

An ability to make parents feel at home and welcome in the center

A sensitivity to the full impact of cultural and socioeconomic factors on the families being served

An understanding of the policies and philosophy of the center

An ability to communicate easily with the staff

An in-depth understanding of the life style and child-rearing practices of the ethnic groups represented

The knowledge, ability, and desire to bring parents into all aspects of the program

An understanding of the differences between individual children, and an ability to observe and evaluate children's behavior

An ability to communicate with children and families in a warm and sensitive way

An ability to accept people as they are, and to listen to and understand their concerns

The ability to express caring for a person or family by concrete actions which alleviate problems

A respect for a person's choice of confidants

An awareness and knowledge of community resources available to all families.

The social worker's major functions in a child-care center are threefold.

CONSULTANT TO TEACHERS

The concern most often raised by teachers is associated with the learning and behavior problems of children. Especially in early childhood programs, the emphasis is on the total development of the child: intellectual, social, physical, nutritional, and emotional. In order to support all phases of development, the social worker tries to stimulate understanding and communication among the child, the teacher, and family members to ensure a regular exchange of information about the child's progress, health, and any current or chronic problems. It is then easier to evaluate the teacher's concern, based on his or her knowledge and impressions of the child and family.

The social worker can observe the child in the classroom and play yard over a period of time in order to allow for the many moods of a child's personality to emerge in a variety of play situations. A discussion with the teachers should follow, regarding the child's interests, achievements, sociability with classmates and adults, the pattern of learning or behavior problems noted, the differences and similarities observed by different teachers, the relationship among family members, and the variety of approaches already attempted by the teachers. Depending on the degree of seriousness of the concern, the teacher already may have discussed it with the parent in a supportive way. When a teacher specialist or a psychologist is part of the center staff or consulting team, he or she can offer valuable insights about possible specific learning problems. Any other impressions or evaluations completed by other support staff, such as the health consultant and speech therapist, should be assessed. The school folder for the child sometimes reveals threads of family history or past school records that are useful. When appropriate, a next step could be a child conference among all staff who are involved, and possibly the parents.

Frequently, a child new to the child-care center is conspicuous by his or her misbehavior. The reasons vary, of course. It may be a result of high expectations or dissension in the home, or a mismatch of personalities between teacher and child, or a parent's "early morning blues," or a child's poor sleeping habits. Or it could be a carryover from sibling relationships. Or it may be simply the child's way of adjusting to an entirely new play situation where learning to share playthings, making new friends, and being away from mother for unusually long hours is upsetting. A different approach by the teacher may be all that the child needs.

Sometimes a parent lets the teacher know he or she is dissatisfied with the school because "no one is teaching my child to read." If the teacher or the administrator does not satisfy his concerns, the parent may be referred to the social worker to talk further about the child. Assuming in this instance that the child is performing satisfactorily in the teacher's view, the social worker may use this conference with the parent to get acquainted and to be supportive of his concerns, to inquire further as to his hopes and expectations for his child, and to determine that the parent's needs supersede those of his child. Several things, concurrently, might follow: the sharing of the child's educational progress more regularly with the parent, continuing observation of the child by the social worker and the teacher as a preparation for another parent conference, and a joint decision by the teacher and the social worker to use an educational measurement check list for charting the child's academic readiness and knowledge.

The role of the social worker is to evaluate the situation that is presented and to be both sensitive and supportive to the family and to the school staff. This situation, as described above, may culminate in the parent enhancing her mothering skills after observing the child at school for several hours. The parent may also be encouraged to join the teachers on some regular basis and teach reading readiness skills, thereby gaining the satisfaction of contributing to the education of the children, even if only for a half-hour each week. The teacher, in turn, may be inspired to create special materials for the children which might contribute to a new activity worthy of sharing with colleagues. However idealistic all of this may seem, it is most likely to follow in the footsteps of parents and staff recognizing and praising each other's efforts. Nobody, amidst those lonely four walls at home or after a long, hard day of teaching, ever receives enough appreciation.

DIRECT AND CONSULTANT SERVICES TO PARENTS, FAMILY, AND COMMUNITY

The social worker should be available to parents as they begin to use the child-care center. Greeting parents and their children during the initial visiting hours provides the opportunity to welcome them at a significant moment, to anticipate their inevitable questions, and to reassure them about the adjustment time needed for parent and child in a new environment. A descriptive brochure, or a printed "Letter to Parents," offered at this time reinforces the personal contact; it also invites a parent to express impulsively any common concern he or she may have about the child. If that concern leads to a longer discussion, or a later appointment, it represents a further step towards active participation in the life of the center at the earliest possible moment.

By taking time to talk informally with parents throughout the day, the social worker has a better chance to become acquainted with the family and to understand its needs. To be a familiar and friendly face before a parent might need advice or counseling is an important part of establishing a constructive relationship. The social worker can attend parent meetings at the center and may organize coffee hours and regular rap sessions in order to be available to families at different times.

Because death in the family, separation of parents, physical violence, mental illness, and other family crises can happen anytime, the availability of the social worker to parents is crucial. The social worker should have a well-established routine in which he or she can be reached throughout the working day at any location by phone. Secretaries must be alerted to the possibility of urgent and emergency calls, and should be able to handle them. Also, the administrator and teaching staff should arrange a means in which emergency calls can reach the social worker during off-duty hours and on weekends.

It is seldom possible to see all parents at one time, even at family-night pot-luck suppers or required parent meetings. Especially at the beginning of the school year or enrollment period, centers should have orientation nights at which the full child-care staff can be introduced to parents. Even if only briefly, it is useful to describe the social services offered by the center and to identify each service by the staff person. A common theme that should be emphasized is neighborhood and community, because the center should represent the best of community life: the worth of family life, its traditions, and its heritage. Materials describing the center can be distributed; a display of community resource brochures would be timely. Because child-care centers vary so widely, a committee of staff and parents together can determine what else might be helpful to new families.

Parent discussion groups offer an excellent means for unifying families under the banner of child-rearing practices. Parents usually feel a sense of relief in sharing their concerns with others and a sure sense of helping others when the experiences of parenting are welcomed. The social worker, other social-service staff, and community-resource people all can be utilized during parent meetings. For the social worker on the child-care staff, parent discussion groups provide the base on which to build child growth and development seminars, parent panels, and the beginning of preventive mental health practices. Full use should be made of child-care medical or nursing consultants, if available, or the visiting nurse association, the public health nurse, well-baby clinics, and the local and county health departments.

The rights of parents and children are respected, and the richness and diversity of ethnic groups are recognized, when the child-care center offers a bilingual social worker or teacher for every language group represented in the community. To further extend this service, the social service staff should provide bilingual staff for home visits, especially to members of families who may feel restricted to their homes for cultural reasons. An arrangement should exist so that bilingual staff members are on call for family crises and emergencies, too. However, staff members must be respectful of any family which may not welcome home visits, planned or unplanned.

Social services to children and families should be built-in to the child-care center from the first inquiry about application and enrollment to the last contact with families as they are transferring, moving, or dropping out (1). The experience of seeking child care may be a difficult and upsetting experience for a family. The adjustment to the rules of the center, the sudden interest of child-care staff in the child and his or her medical, developmental, and family history, and the expectations that the staff has of the child and family may seem rather demanding. It should be balanced by the sensitivity of the social services staff in the enrollment process and their regard for the feelings of parents and the importance of privacy at this time.

Other social services should include the examination of alternative plans for the care of children; the inclusion of parents in fee-setting; the explanation of what parents can expect from the child-care center in meal-planning, arrangements for medical emergencies, discipline, health and safety measures, and procedures for getting information, making suggestions, and discussing grievances (2). Social services staff can assist families in recognizing the need for community resources and by providing information about what the agencies offer and how they can be contacted. When a family wants help in the referral itself, the social services staff may be directly involved in initiating a phone call on the family's behalf or in a follow-up contact. It places a clear responsibility on the social services staff to be as informed as possible about the eligibility requirements of community agencies.

Consultant to Child-Care Center Administrators and Staff

The social worker must identify with all aspects of the child-care center. As a consultant, he or she has a stake in the success of administrative leadership, the evolution of staff leadership, staff morale and issues of importance to staff, the excellence of food services and nutrition, the incorporation of enlightened in-service staff training, and a balanced, comprehensive program that meets the needs of children and families. This point of view suggests that the social worker wants all the same things a dedicated administrator wants in a child-care center—and, most definitely, this is true. Just as an administrator must decide how to insure that the teachers meet and befriend every new parent individually in order to establish a positive beginning relationship, the social worker picks up the pieces of misunderstanding throughout the year when that same parent-teacher relationship does not exist. Carried out to its fullest extent, the social worker must contribute remedies and services to fill unmet needs of children, parents, and teachers whenever the program does not already provide for them, or when there is a breakdown of services or relationships.

As a consultant, the social worker may be meeting with a head teacher and staff on a weekly basis, trying to resolve the many challenges and problems which are bound to arise. Issues, for example, which are common to all child-care centers are: practices of involving and relating to parents, the consistent care and handling of children, staff relationships with children about sex play and "bad" language, lunch table rules and expectations, and personnel problems among staff members.

At one time or another, the social worker may have reason to play the role of devil's advocate, change agent, enabler, conciliator, negotiator, liaison person, and expediter. There is an implied commitment to working within the system which exists, but also to attempt to change that system when it does not meet the needs of children and families. At a time when a program is not functionally adequate, or there are gaps in supervision, leadership, or services, the social worker has the responsibility of finding a remedy and seeing that the needs are met.

CHILD CARE IN BERKELEY

There are a wide variety of child-care programs, public and private, in the city of Berkeley. Unfortunately, there are not enough to go around. Many more children and families could benefit from an expansion of programs of all kinds and an even more varied offering of programs and services. For example, some parents desire more comprehensive day-care services, such as drop-in 24-hour child care, infant care, and free child care for all parents who are students or working.

The primary attempt to remedy this situation is being made at the city government level. Several years ago, the Berkeley City Council funded a survey of the child-care needs of families throughout the entire community. This resulted in the creation of the Berkeley Child Care Development Council with a full-time salaried staff and director. This office has become a resource to the community as a clearinghouse for public and private child-care facilities, and as a developer of quality child-care programs for the community. A citizen's council of parents, staff, and community resource representatives, including the public schools, meets regularly as an advisory body.

A recent Berkeley Child-Care Development Council project, established in January, 1974, is the Caring Center, which is a day-care center for neurologically and autistically handicapped children. The center is jointly funded by the city of Berkeley and the state of California. It serves children from ages five to twelve years before and after school, opening at 7 A.M. and closing at 6 P.M. This is the first time that the state of California has released money for this specific type of facility.

Another Berkeley Child Care Development Council program is for the care of sick children at home, where working or student parents simply phone a 24-hour answering service, and arrangements are made on a sliding-scale fee basis, depending on family income.

Other groups, organizations, and educational institutions in Berkeley have created ongoing programs on a part-time or full-time basis. The University of California, Berkeley campus, offers six day-care centers for children of students, including age groups from infancy to six years of age. The YWCA provides a drop-in child-care program daily for a small fee; in addition, counseling is available to mothers at the same location. Another YWCA child-care resource, called "Bananas," serves two purposes: first, it acts as a clearinghouse for setting up voluntary neighborhood play groups for young children where adults make their own arrangements and no money is exchanged; second, it is also a referral source for parents who need any and all kinds of child care services in YWCA-approved private facilities.

The Berkeley public schools offer a diversified assortment of child-care programs throughout the community, serving some age groups from infancy to eleven years. For example, the Berkeley High School Parent-Child Education Program was established three years ago in order that high school students who are also mothers might continue their education. They receive credit for two required courses in child development, in addition to their regular high school courses. Their children are accepted for child care at the center from the ages of three weeks to two and a half years for the entire school day.

Extended day-care programs exist in nine public schools for children in kindergarten to sixth grade whose parents are students or working. They provide before- and after-school supervised activities.

For children under kindergarten age, there are three types of child-care programs:

1. The children's centers, which provide educational day care for children (ages two, three, and four) of student or working parents

2. The parent nurseries, three-and-a-half-hour pre-school classes requiring parent participation, for children ages two, three, and four

3. The early learning centers, which combine the above programs with primary school, enrolling children from two and a half to nine years of age in cross-age groupings.

A Department of Early Childhood Education was created in 1965. Although parent nursery and children's-center programs have been in existence for many years in the Berkeley public schools, this was the first time they gained departmental status.

Due to the multiethnic population of Berkeley, all public school classes are balanced according to the age, race, and sex of the children. Black children and white children are almost equal in number, and they comprise more than 90 percent of the total enrollment. Other ethnic groups represented are: Japanese, Chinese, Filipinos, Chicanos, and American Indians.

CONCLUDING REMARKS

The intent of this chapter is to illustrate how a social worker may fulfill a variety of roles which are supportive, conciliatory, collaborative, cooperative, and evaluative in nature. The common thread that unites them is one of enhancement of the family and each of its members, the vital importance of parenting, the uniqueness of each child and his or her individual differences, the commitment to positive relationships within the family structure, and the promotion of the basic problems of normal child growth and development.

NOTES

1. Day Care Workshop. Chapter VII: Social Services. Washington: Office of Child Development, Department of Health, Education, and Welfare, 1970. Pp. 85–88.
2. *Ibid*.

Chapter Twelve

THE IMPLICATIONS OF PIAGET'S THEORY FOR DAY-CARE EDUCATION

Keith R. Alward

All signs indicate that day care is coming to the United States and that in the next few years we will witness a dramatic increase in facilities for the care of young children. As day care presently is conceived, this care will be delivered outside the context of the child's home environment and outside the conventional mother-child relationship. Most likely the social effects of day care will be diverse and significant. Changes in family living patterns, as well as the family structure itself, employment patterns, the economic structure, and educational needs and priorities, are all forseeable consequences of extending our commitment to day care. However, even more important than its anticipated effects is the responsibility associated with this commitment.

Whoever assumes the task of planning or implementing day care must also assume a responsibility that goes far beyond that presently associated with elementary education, one that is directly traceable to the age of the day-care child and the extent of his contact with the day-care setting. As opposed to the elementary age child who, at the youngest, is generally five or six years of age, and who spends an average of five to six hours a day in school for nine months of the year, the day-care child may be two months of age and spend nine hours a day, twelve months of the year in a day-care setting. The elementary school child may spend 12 per cent of the year in school, whereas the day-care child will spend approximately 27 per cent of the same growth period in the day-care center. The implications are enormous. At an extreme, day care may well mean that a large segment of the early childhood population will be raised virtually outside the social-psychological patterns that operate in the home environment. Not only in terms of the amount of contact, but more importantly the age of the child, day care can be viewed as having a very significant impact on the child's development. Depending upon the form of experience offered, day care could well constitute a source of cultural, social, emotional, intellectual, and physical deprivation. The seriousness of this risk will be a direct function of the degree to which the day-care setting provides an extension and continuation of the home environment and the extent to which the experiences offered provide adequate stimulation for the normal development process. It is in these areas that the responsibility of day care must be met.

I'd like to put this issue into the context of educational objectives and their place in day care. I think we need to understand that many view day care not only as a service for the working mother, but also as a means of breaking the poverty cycle: freeing low-income mothers for work, providing a job market for low-income mothers as day-care workers, and providing an early educational experience for children of low-income parents. There is an implicit assumption that poverty is a consequence of the individual's failure to secure an adequate job and the failure of the

individual's social-cultural milieu to provide the appropriate preschool experience for later educational success. In
my opinion, this assumption is unjustified and wholly inadequate as an explanation of the social dynamics that contribute to an uneven distribution of wealth. However,
rather than pursue the basis for this opinion, I simply want
to suggest that the very nature of day care and its anticipated effects is such that it will result in an enormous effort
to make day care an educational experience for the children
of low-income as well as non-low-income parents.

I do not wish to suggest that day care should not be
involved in educating children. On the contrary, in the
child's first four years he accomplishes a Herculean task of
acquiring knowledge, skills, values, and attitudes, and in
the sense that education is generally directed towards these
ends, it's fair to say that day-care age children are normally
involved in an educational enterprise that should be supported. However, we need to take a careful look at what
education means at this age. While many continue to view
all education as a process by which social knowledge (that
is, names for things, verbal concepts, mutually shared ways
of doing things, and so on) is taught to the child, the majority of knowledge acquired by the preschool child actually is
minimally social and not learned through what we generally regard as the "teaching process." I will try to clarify this
distinction by reference to the works of an important child
psychologist, Jean Piaget. What I'll try to show is the nature
of knowledge which the preschool child acquires, the manner of its acquisition, and how an educational environment
can support the developmental process.

PIAGET'S THEORY OF KNOWLEDGE

I have referred to knowledge being acquired, and this may
be somewhat misleading. The thrust of Piaget's theory is

that knowledge, even perceptual knowledge, is not passively taken from the environment. Knowledge is not something which we consume or acquire from someone or something, like acquiring a car or a new dress. Knowledge is something which is constructed, made up, and put together by the owner. It is never something which is abstracted in toto from the environment, no matter what the nature of that environment might be. Knowledge is always a constructive process that relates to the activity of the knower. And the nature of these activities is of key interest to us as educators because we want to know what the child should do in order to learn the skills and "acquire" the knowledge that the world will expect of him. The notion of action is important not only to the issue of what educators should ask children to do, but to the whole issue of how "doing" or "acting" relates to learning and knowledge.

In a sense we come to know our world by acting upon it in increasingly complex and coordinating ways. Knowledge increases as we come to understand the relationship between our various acts and their relationship to outcomes. We grow mentally as we begin to combine old acts in new ways to reach new goals or to construct new acts to realize old goals in new ways.

It is understandable if you find it difficult to grasp the notion that knowledge is not acquired but constructed and that actions are intimately tied to this process. You may ask, what acts are we talking about? Or you may ask, in what ways am I acting or constructing my knowledge as I read this paper and acquire new insights? Clearly some acts, like moving your eyes as you read, are involved here; but they bear little obvious relation to the process you use to get meaning out of the words. Obviously there are some types of acts involved, acts such as making judgments, modifying judgments, connecting thoughts together, separating ideas that look the same but turn out to be different, putting ideas into order, relating one idea to another, and so on.

We are not used to calling these processes "acts," but in Piaget's theory they are acts—which if you prefer, we can call "mental acts." But when you were an infant, you did not know your world through mental acts but through actual motor acts of grasping, putting into the mouth, looking towards something heard; and it is these kinds of acts that lead to growth of later mental acts. Central to Piaget's theory is the idea that knowledge is constructed and that the constructive process is an active process on the part of the learner.

THE STAGE THEORY OF DEVELOPMENT

An immense amount of empirical data has been gathered over the last 50 years by Piaget and those associated with him and by countless others who have pursued the implicatins of his findings. The overwhelming evidence is that the child progresses through a number of definable stages characterized by the types of actions which the child can carry out, and hence the kind of knowledge which the child either can construct or understand. It is common knowledge that the child fails to grasp many adult concepts or that the child's conceptions of reality are limited and often contradictory or incomplete. But until recently, we've viewed this characteristic of the child as reflecting a lack of experience, or an inadequate exposure to adult thinking. Certainly we've recognized that a child learns only so fast and that preliminary skills and concepts are necessary for a foundation for later thinking and more complex skills. But our view of this generally has been that once the child can use language we can teach him anything that can be expressed with language and once he can read and write, the child has access to all knowledge that exists in written form.

The idea was that the child functioned like the adult and that educating him was simply a matter of filling him with adult thought. Piaget has changed all this. The child is not simply a little adult, an adult with less knowledge. A child structures and organizes his knowledge about the world in a way that is peculiar to the kind of activities which he can coordinate and this is qualitatively different, different in kind, from the way an adult organizes and structures his world. (1) And of course, it is not just an issue of children versus adults. The child of 13 has different capacities from the child of seven, the child of seven is a different kind of mental organism from the child of four, and so on. Today the notion that there are different stages in the child's development is a clearly documented position. The last 50 years have seen an enormous effort to show that the position was wrong, which is one of the ways theories are scientifically tested. However, all evidence points to the validity of a stage theory of development.

Let's take a closer look at this stage theory.

First, it says that the child's mental development consists of a number of stages and that each stage is characterized by the type of knowledge which the child can construct; that is, the type of actions which he can generate and coordinate.

Secondly, the sequence of these stages is invariable—that is, they always unfold in the same order. A later stage never shows up before a stage which is structurally earlier in the sequence.

Third, the development of these stages has a general, and not a specific, relationship to age. For example, the first signs of a stage called "concrete operational thought" are seen generally, but not necessarily, around seven years of age.

Fourth, these stages do not unfold as a function of strictly physiological development; that is, you cannot physically support a child and isolate him from an active interaction with the environment and expect to see mental development. The theory of stages is a theory of the organism's active interaction with a varied and simulating environment. But it is not a theory that says that the child passively copies reality by experiencing it. It is an interactional theory that holds that the child brings his capacities to bear on the environment and in so doing, they become modified.

Fifth, each new stage arises from a modification of the previous stage and this takes place only if the structures responsible for the actions of the previous stage are functionally exercised; that is, used in a context appropriate to their function.

Sixth, and possibly most significant to educators, no effort that we have yet been able to bring to bear has the effect of altering the natural sequence of this development. That is, no matter what we have tried to do—and all of our presently best ideas as to what to do have been tried—a particular stage of knowing cannot exist unless the previous stage of knowing specified by the theory has been developed and functionally used by the child.

Likewise, our best efforts to speed this process have failed to produce any gains which would justify the effort. When these gains have been achieved they are usually in the order of a mean gain of less than six months and they usually have been obtained in an environment which simply affords greater opportunity for the child to explore the implications of his own actions.

I've emphasized that knowledge is constructed or made up and that what the child can be expected to know or understand will be a function of his capacity to coordinate previous knowledge and actions. This capacity increases in definable stages—each stage being characterized by the nature of possible actions and their coordination. Furthermore, each new stage is achieved through an appli-

cation of the actions made possible by the preceding stage
—an application that leads to new coordinations of previous actions and thus, the possibility of new actions themselves.

It might be helpful if we look at some of the earlier
stages and their broad characteristics. However, first we
should consider some of the implications of a stage theory.
If, for example, the stages unfold in the order A, B, C, D,
E, then at least two points follow. One is that if a later stage
(for example, E) is characterized by the inability to construct certain knowledge, then it follows that the same failure will be observed at earlier stages (D, C, B, A). The
second is the converse, that if certain knowledge can be
constructed at an earlier stage (for example, A), then it can
also be constructed at later stages (B, C, D, E). However,
if something can be accomplished at a later stage, it does
not necessarily follow that it must also be possible at an
earlier stage. With these points in mind, let's look at some
of the early childhood stages and some of their characteristics.

THE SENSORY-MOTOR PERIOD OF CHILD DEVELOPMENT

The sensory-motor period is the first broad developmental
period and generally covers the first two years of the child's
life. However, as is true of all the stages, they are not
directly related to age and thus some children may move
out of the sensory-motor period prior to two years of age
and some at a later age. In general, the end of the sensory-
motor period is marked by the beginning use of language
and representative symbols—somewhere around 18 to 24
months of age. Up until this time, the child's knowledge is
directly related to his own physical actions and not to symbolic representations of the real world.

In Piaget's theory, each action that the child produces

is directed by a scheme or a plan for action. As a contemporary analogy, you might think of a scheme as a computer program that underlines specific actions or functions performed by the computer. All mental development is seen as the elaboration of schemes and the evolution of their coordination. During the sensory-motor period, these schemes relate to physical (sensory-motor) actions and not to mental ones—as when we imagine a particular sequence of actions or when we make deductive judgments.

There are six distinct developmental stages within the sensory-motor period. I will give a very brief characterization of each stage and then give some idea of the kind of knowledge which is constructed during this two-year span.

The first stage covers approximately the first month of life and consists of the child's reflexive responses to environmental stimuli. The nature of adaptation to the external world or what the child can accomplish in that world, is another way of looking at the kind of knowledge that the child constructs. At this stage very little adaptation or accommodation takes place. Such feats as being able to turn in the direction of interesting stimuli, locating the nipple when feeding, and so on, constitute the height of the child's achievements during this period.

The second stage, lasting from approximately one to four months of age, is characterized by the infant's coordination of his own actions carried out on himself. Coordinating movements of one part of the body with those of another—for example, coordinating hand-eye movements in such an exercise as putting the hands together—characterizes this period. In the first few months of life, the infant is not yet interacting in a manipulative and directive fashion with the external world. Seeking out and grasping objects, manipulating objects to create interesting effects, and so on, are coordinations reserved for the following stage.

Piaget regards the third stage, covering the period from four to eight months of age, as the beginnings of mental life, because it is here that we first see the child accommodating his actions to affect a reality outside of himself. This change in the child's sphere of influence is accompanied by the beginnings of intention—the desire to create effects in the external world. Acts such as grasping and bringing objects to the mouth, shaking objects, and pursuing activities to recreate previously interesting events characterize this stage. At this time the child begins a first distinction between means and ends. For example, if an interesting event occurs in conjunction with some activity on the child's part, he may repeat the activity in order to cause the reappearence of the interesting event. This rudimentary behavior indicates a change in the child's activity in order to create an event outside of his previous control, and hence a beginning construction of a desired end and a means of achieving it.

During the fourth stage, which ends at approximately 12 months of age, the child makes rapid progress in modifying actions to achieve new ends. As the child attempts to reproduce certain consequences in the external world, he encounters obstacles. The encountering of obstacles helps the child to differentiate or distinguish between actions that appear to be the same but have different effects as well as to discover that different kinds of acts can have the same effects. With this progress the child becomes increasingly facile in both the generation of goals as well as the means of their achievement.

During the fifth stage, lasting until approximately 18 months, the child begins to combine actions to form complex systems for achieving goals. Of course, along with this new power comes an increased capacity for defining the goals themselves. The child of this period engages in a constant process of experimentation, slightly altering actions to investigate slight variations in outcomes. For example, one of the landmarks of the fifth stage is throwing objects. Many a distressed parent will agree that throwing is not a new occurrence. But here we find that the child is actively experimenting with throwing; he's become interested in the path of the object and its relation to how it was thrown, whereas before he was interested only in the spectacle itself. This experimentation in means expresses itself in imitation as well. If the adult is modeling a behavior which the child has not seen himself make—such as closing the eyes—he will imitate the act by successively closer approximations: he may first close his mouth and then his eyes. Related to this is the child's search for the cause of something that has caught his attention. He may actively manipulate and explore until the result is again achieved. Another characteristic of this stage is the child's use of pointing to indicate or make reference, thereby indicating a primitive beginning of symbolic reference. The ability to imitate also reflects a representative function; but like reference through pointing, the child cannot yet imitate or represent actions that are not in the present. However, objects which are no longer present still exert some control. He may look for an object even though it has not been present in his ongoing activities. This also illustrates an extension of the child's use of time. Time now exists in more than just the immediate spatial-temporal realm.

It might be illustrative to explore some of the things that the child can not yet do at this fifth stage of the sensory motor period. For one, the child's coordination of spatial displacements (movement of objects in three-dimensional space) is still rudimentary. For example, if a ball that has the child's interest is moved along a path, obscured at various points, and finally hidden at the end of the path, we might expect a child quickly to locate the object. However, the child in the fifth stage will be stumped by this problem even though during the next stage it is easily solved (see Fig. 12–1).

Another aspect of the child's capacities at the fifth stage is illustrated in trying to put a string of beads into a

Fig. 12-1

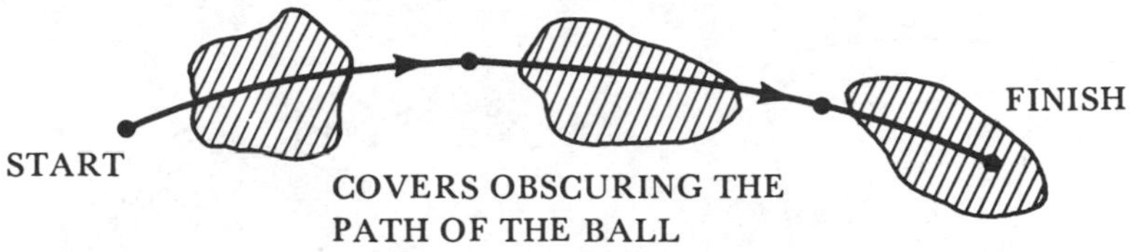

START

COVERS OBSCURING THE
PATH OF THE BALL

FINISH

Situation: The child is engaged in play with an object that is inter-
                      esting him. In this example, the adult shows the child
                      that he has the ball, and then, hiding it in his hand
                      (but not letting his hand leave the child's field of vision),
                      the adult passes his hand under three covers leaving the
                      ball under one of the covers. The adult then shows the
                      child that the ball has "disappeared"—that it is not in
                      his hand. The child is encouraged to locate the ball.

Behavior of a Child in the Fifth Stage of Sensory-Motor Development:

                      Typically, the child looks under the first cover, possibly
                      the second. If the ball is under the third cover the child
                      will have missed it and begin looking around the room
                      as if the ball were elsewhere than under the covers. The
                      child rapidly looses interest in the ball's whereabouts.
                      The ball is seldom found.

Behavior of a Child in the Sixth Stage of Sensory-Motor Development:

                      The child goes immediately to the third cover. If the
                      ball is not found, the remaining covers will be searched.
                      The ball is always found.

drinking glass. Before 18 months, on the average, the child
will try to put individual beads into the glass with the inevitable results that either the glass is upset or the beads
slither out into the floor. At the sixth stage, the child will
ball up the beads and place the whole ball into the glass in
one smooth action (see Fig. 12-2).

Both of these examples relate to the development of
a capacity to represent reality internally. When this capacity
emerges we have the beginnings of symbolic behavior, such
as language, which results from actions which have become

Fig. 12-2

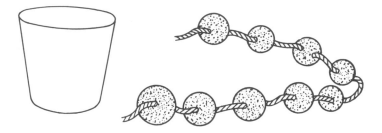

Situation: The child is engaged in playing with a number of
                  objects including a drinking glass and a string of beads.
                  The child is trying to put the beads in the cup.

Behavior of a Child in the Fifth Stage of Sensory-Motor Development:

                  The child puts one end of the string into the cup and
                  then picks up another portion of the bead string. The
                  result is that the end inserted into the cup slips out or
                  the cup tips over.

Behavior of Child in the Sixth Stage of Sensory-Motor Development:

                  The child picks up the beads and crumples them into a
                  ball in his hand and places the whole ball into the cup,
                  thus succeeding.

internally represented. In the case of the ball, the child fails
because he does not have a mental image of the movements
of the ball and their relationship to space. In the container
and beads situation, the child cannot internally visualize or
anticipate the behavior of the string of beads as he tries to
put it into the drinking glass.

Piaget contends that prior to the sixth stage, at which
time actions begin to be internalized, the child does not
have the capacity to represent reality. Up to this period, all
reality is represented through external actions in which
objects are recognized in terms of actions related to the
objects. After this period, objects can be represented internally, without external actions carried out.

This new step forward marks progress in the child's understanding. At the sixth stage (occupying the period from approximately 18 to 24 months), we witness fairly stable concepts of objects, causality, time, spatial displacements, and numerous other categories of knowledge. (2,3,4) However, these concepts are stable only within a very specified context—that of the child's actions—and while the development of this knowledge has yielded the capacity to represent knowledge symbolically and internally, a whole new developmental progression is in store before these same areas of knowledge will be stably applied in a symbolic manner. In fact, the full course of the mental changes already in progress will not be completed for approximately another 14 years.

REVERSIBILITY AND THE STABILITY OF THE CHILD'S THOUGHT

While recognizing that the preceding description of sensory-motor development is indeed sketchy, I want to emphasize that even in the child's first two years an enormous construction of knowledge has taken place; knowledge that shows itself in the child's increased capacity to organize his own actions in a stable relationship to the world of physical events. If a ball rolls under the couch, the two year old can retrieve it by following its path; the one year old, on the other hand, acts as if the ball has disappeared from reality. Further examples of sensory-motor knowledge are no less numerous than the child's capacity to formulate and act upon his own intentions.

It's important to realize that sensory-motor knowledge and, in fact, all knowledge, is not a biological given which simply unfolds with time, nor is it something that is simply taught. Knowledge is constructed by coordinating actions carried out on the real world and the increasing complexity of these coordinations characterizes mental development.

In a rough sort of way, the preceding description of sensory-motor stages indicates a developmental progression from simple reflex responses to stimuli, coordination of body actions leading to coordination of body actions in conjunction with the movements of objects, leading to combining a number of sub "means-ends" schemes to reach goals, leading to a coordinated system of "presentational images" which guide the course of schemes being coordinated as both means and ends, leading to an internalization (or representative imagery) of the preceding coordinations.

At the end of the sensory-motor period (approximately twenty-four months), the child's thought is characterized by what Piaget calls "reversibility." (5) The concept of reversibility is central to Piaget's theory and important to understanding all of mental development. Reversibility characterizes the state towards which all coordination (and hence, development) tends and underlies such important functions as symbolic representation, imagery, and logical thought. The emergence of reversibility also underlines the stability of thought.

By "stability," Piaget means that the child's knowledge has evolved to the point where he has fairly predictable and reproducible interactions with the world. Let's look at this notion a little closer. Various actions that the child can perform on the world come to be coordinated into systems such that any act that can be carried out in the system can also be undone or reversed. While this seems quite simple in principle, the notion of reversibility is central to all systems that can predict reality. Even the most complex predictive system, mathematics, is dependent upon reversibility, and Piaget has gone so far as to suggest that mathematics and logic have developed through the history of man's reflection upon the reversible properties of his own thought; that mathematics and logic are formalized reflections of the way the mind operates. (5, 6) This notion is receiving wide support in modern mathematics.

While the end of the sensory-motor period is characterized by a reversible system of actions—which means that there is a stability to the child's thought—the reversibility only applies within a limited sphere. That is, while the child is beginning to have a predictable (reversible) relationship to his immediate actions on an immediately affectable outside world, he is not yet capable of a reliable understanding of the outside world upon which he has no direct relationship. This limitation also extends to symbolic representations of reality which the child is just beginning to construct.

By way of example, consider the child's capacity to organize spatial displacements as illustrated in Fig. 12–1. The child can keep track of an object in this as well as more complex hide-and-seek games, and this indicates a logical organization and understanding of the relative movements of objects—an understanding capable of generating deductions and inferences (predictions).

In contrast, we might consider the behavior of a child at a much later age, around four or five years old. If, for example, the child is traveling in a car with the moon always in view, he may well exclaim that the moon is following him. If we ask him whether it follows others as well, he may say yes or no. But in either case, he does not realize the implications of his assertion; that if it follows others and himself simultaneously, there must be many moons, or that if it only follows him, then the moon must have a special relationship to him that it does not share with others. We call this thought incorrect, but when we understand the structures of the child's mind during this later period (called the "preoperational period"), we come to realize that these are reasonable outcomes.

While this is a limited example, it illustrates the fact that the structures of sensory-motor reasoning are not adequate to organize such distant phenomenon as the relative movements of the moon. By "distant" I do not necessarily mean physical distance like miles, but rather something like abstractness. In this case the abstractness is partially related to the fact that the movements of the moon cannot be directly manipulated and organized by the child. But abstractness also applies in some instance where direct manipulation is involved.

For example, quantitative notions such as weight, numbers, length, amounts, and so on, are all abstractions. There are no numbers in nature. Rather, the abstraction of quantitative concepts grows out of coordinating our actions on objects. The child's comprehension of these abstractions occurs much later than previously assumed. The preoperational child (between approximately two and seven years of age) assumes that the number of objects changes when the objects are rearranged, that the length of an object changes with change of position, that the amount of a substance changes with changes in shape, that the weight of an object changes with changes in shape, and so on. In any area of quantification we find a lack of stability in the preoperational child's thought. (7, 8, 1)

This same lack of stability is evident in other areas as well, including the organization of space. When I was a little over five years old, we moved to a new home located a hundred yards from a deep ravine running from the mountains into the valley. I was fascinated by the "big hole," and although, when I was at its edge, I could clearly see that it extended in a long straight line, it was some time before I could actually navigate from the house to its side. I frequently asked a neighbor to take me to its edge, not realizing that if I were to travel in its general direction I was bound to reach my goal. I first made the trip on my own by following a specific path through the brush and only later did it dawn on me that any number of paths lead to its edge.

If we ask a preoperational child if he has a brother and he says yes and we ask if his brother has a brother, he's quite likely to say no. Here again we see a limitation in the predictive or logical power of the child's thought. He does not see the fact that he has a brother also implies that his brother has a brother. One does not yet follow in a necessary manner from the other.

As one last example, consider the questions: Are there more children in the world or more little boys? Or: Are there more roses in the world or more flowers? The young child does not answer these questions correctly even if the questions are put into the context of actual physical materials that the child is playing with; for example: Are there more wooden blocks or blue wooden ones? Many adults are surprised that such simple and commonsense questions are missed. However, if you consider the question: Are there more animals in the world that are not birds or more that do not fly? you get some idea of the problem faced by a child with simpler versions of "class inclusion" problems (9).

You can probably anticipate the next episode of this story, that somewhere between seven and twelve years of age, the child no longer thinks that the moon follows him, nor that changes in position and shape affect quantity. He realizes that his brother has at least one brother (namely, himself), and that the members of a superordinate class are more numerous than those of a subordinate class. The big question is, how is this progress made? Because of limited space I have not given you the appropriate data for drawing your own conclusions. The examples I've given do not adequately characterize the actual manner in which the child's knowledge is investigated, nor do the examples adequately represent all areas of knowledge that have been studied by Piaget and others (10, 11, 12, 13, 14, 15). Thus you might feel that the examples are more easily or better explained by factors other than "a reversible organization of internal mental acts." You might suggest that the "concepts" haven't been taught to the child, or that he hasn't been exposed to the right experiences. Or, you might suggest that it is a language problem; that possibly the children didn't understand the words or instructions or the questions. Another possibility is that the described behavior applies to some children but not to others, that maturation simply produces this form of thought with time. All of these are possibilities, and a good deal of research by Piaget and countless others has gone into their examination. So far Piaget's theory has provided the most comprehensive explanation, although there are some promising elaborations on Piaget's theory presently under consideration. If it could be shown that other factors such as social learning, conditioning, or language constitute the underlying dynamics of mental development, it would be a great service to psychology. But the point is, as yet nothing of this sort has been shown.

THE DEVELOPMENT OF INTELLIGENCE

According to Piaget, the answer to the question, How is mental progress made? is found in understanding how intelligence itself develops, or broader still, how any human function grows more sophisticated through its application (5). Let's try to use, then, the concept of a developing intelligence and hold in abeyance our associations with tests, individual differences in abilities, and test scores designed to characterize the child. If we can determine how the child's intelligence naturally develops, we shall be in a much stronger position to talk about what educators should do and how they can realize their goals. Actually, we've already covered a number of basic concepts regarding the development of intelligence.

First, the development of intelligence always is based upon actions, just as the evolution of life is based upon living. In the first two years of life, the child's actions are strictly motor and perceptual. These actions are not yet mental in the sense that they are for older children. Everything the child knows is expressed in his physical activities. Here we find the precursors to later mental acts. For example, when the child puts a wide assortment of objects into his mouth—we hope not all at the same time—he has formed a general concept or class of objects, and one of the ways that he knows or understands these objects is that they are things that fit into the mouth. When the child carries out a sequence of acts that are bound together, for example, grabbing a rattle to shake it, he must first get to the rattle, then shake it; and here we see the beginnings of serial order. The examples of early precursors to later mental actions are endless. The point is that mental development starts with activity, and all development is a modification of these activities through various stages. As activities become internalized—that is, represented by images, symbols, and social signs—they can become coordinated in new ways, ways that we call "thinking"; but they are still actions that are coordinated with each other.

Second, all actions are functions that change one situation into another; and as functions, actions serve an adaptive purpose in the interaction between the organism and the environment. We've already seen how the first two years of life can be fairly well understood in terms of the child's capacity to construct goals or intentions and to construct means of reaching or satisfying them. This same theme is evident throughout mental development and is a critical element in the "motor" behind this development. Many view the child's growing capacity to solve problems as the main theme in the development of intelligence. But this is only one side of the picture. The capacity to solve problems grows out of the capacity to define problems and vice versa. That is, defining and solving problems are twin functions, and the history of the child's mental development, as well as that of man in general, reflects this twin process.

Third, a good portion (and possibly all) of our actions, whether physical or mental, take the form of mathematical actions: combining (+), separating (–), repeating (X), parceling out cause and effects (÷), comparing (=,<,>, ≈, ≠) and ordering (A comes before B; A is less than B). There are additional actions that we generally do not think of as mathematical, although they are; "going into-coming out of," "getting close to-leaving," "closing-opening," "straightening-bending," and "thickening-thinning" are examples. The point to be remembered is that these apparently simple acts constitute the fabric of intelligence, and the development of intelligence consists of more elaborate and complex forms and coordinations of these actions. Mathematics itself represents a special and symbolically formalized form of these actions and their coordinations.

Fourth, a very significant premise in the development of intelligence is that these actions described above tend to organize themselves as a consequence of their being used. We know that the coordination or organization of these actions becomes more complex, logical, and internally consistent—never less so. That is, there is a constant movement towards more stable and predictable relationships between the child's thought about the world and what actually happens in the world. The degree of this stability is, in large measure, a function of whether the actions are organized into reversible systems. For example, in arithmetic, addition and subtraction are bound together into a reversible system. By adding 5 and 7 to get 12, we realize that taking 7 from 12 will give us 5. The reverse holds as well. The preoperational child does not organize combining (+) and separating (–) into a reversible system and the fact that he eventually develops to a stage where he does seems to have no demonstrable relationship to educational efforts.

Last, in order fully to appreciate the statement that "through their being used or exercised, actions tend to organize in more complex, logical, and consistent systems," you must also appreciate the full meaning of "exercise." Piaget's theory states that development cannot take place unless the systems of coordinated actions are exercised; that is, applied on the real world by the child.

This last point is a key issue. What kinds of exercise should the mind have in order for it to develop? Every structure or scheme underlying the mind's actions has a particular level of reality to which it can apply. That is, it can take in, act on, or assimilate certain aspects of reality and not others. If a structure is applied over and over in the identical circumstance, with nothing changing, the structure will not undergo any adaptation, and hence there will be no further development of the structure. On the other hand, if reality is presented to the child in such a way that it cannot be assimilated, that is, inaccessible to his power to recognize, know, or understand, then again, there will be no development or adaptation, since the structure is inactive and not being exercised. The proper context for exercising the mind is one in which the child is dealing with the familiar, but in slightly varied and new ways.

Consider the structures underlying the child's ability to classify objects and to coordinate his classifications. If he is presented with the task of sorting objects on the basis of color and never involved in classifying on any other basis —a highly unlikely situation—his classificatory structures will not undergo development. However, if the child classifies the same from a number of different perspectives— size, shape, texture, and so on—one can expect some modification of the structures underlying these activities.

Fortunately for all, the mind is superbly suited to searching out those circumstances which lead towards its development. The child is interested only in those things that are familiar, yet in some way dissimilar. The "same old thing" loses his interest; something too foreign never attracts it. But if the child has some way of applying himself, he will search out new possibilities that arise in the context of his actions.

Each act of knowing leads to further questions. Both the realization of these new possibilities (questions) and the means of their solution or answer derive from the same structures. That is, the recognition of a problem is possible only if structures are present which provide some means of grappling with the problem. This may seem surprising to us, since we all recognize problems which we find difficult to solve. Certainly there are many variables, such as the amount of time we have to spend on their solution, and so on. But as we come to define the problem in clearer and clearer terms, we also come closer and closer to arriving at a solution. The point is, the mind regulates itself and grows, and it does this by applying its capabilities in such a way that new obstacles to understanding are encountered, specified, and worked on by the capacities which recognized the problem in the first place. Thus, problem-solving is a natural consequence of exercising functions which the child can already perform.

It is important to emphasize that, in general, the development of intelligence is self-regulating and moves consistently towards higher and higher levels of adaptive knowledge about the world. But it is also important to note that this takes place only if the actual structures of intelligence are exercised, and this means that the child must be in a position to explore the relationships between his actions and their implications. If the child were to function only at the verbal level or the level of rote memory of "facts" or concepts, his intelligence would not flourish. Similarly, if the child were to function only in those realms which never challenged him or in those for which he had no notions of how to act, he again would not develop his mental capacities.

The Universality of Adaptive Intelligence

Fifty years of research have strongly suggested that these propositions about mental development hold for all children no matter where or under what circumstances they live. The process of adaptive intelligence is an integral component of the definition of man—not some men, but man as a biological entity. Furthermore, the products of this process, the child's knowledge, show a parallel form in children from all areas of the world. By "form" we do not mean the specific knowledge or actual experiences, which do vary by culture, but the structures of the thought. There are definite levels of mental organization and all children who arrive at the concrete operational stage, for example —and all eventually arrive at this level of thought—demonstrate certain logical characteristics in their thinking. Operational thought is defined by particular and universal structural properties. This is equally true of all the stages. Furthermore, education in its general sense of "teaching" has no relationship to or effect upon the development of these properties.

We should hope that all who come to recognize the validity of this viewpoint can also recognize its implications for an educational program such as day care, which, on the whole, will serve children who come from a different cultural background than those who monitor, administer, and implement the program. There is diversity in the life styles of the Black, Chicano, Native American, and Asian children; diversity between various members of the same ethnic groups and between those and middle-class white America. And of course, diversity in life style is not restricted to ethnic groups alone but is a characteristic of all groups of people who live and organize their world in different ways.

No basis exists for believing that children who come from different backgrounds, who deal with different experiences from ours or in any way are dissimilar to us, are emotionally or intellectually deprived or less complex in their organization of experience. The notion that difference equates deficit is a myth, supported only by our ethnocentricity and by our inability to understand the fullness of that with which we are unfamiliar. There are many models of man, his relationship to himself, and the fabric of his social life that attest to the realization that the differences in man are a tribute to the varied and unique ways in which man adapts and survives and not an indication of the superiority of any life style or culture over another. But Piaget's theory of development forces this realization upon us in a most direct and empirical fashion, for the theory and its attendant research strongly suggest that all mental activity leads towards development, and all development moves towards a more refined understanding of the experiential milieu in which that development takes place.

While part of the national interest in day care may come from a belief that the children of low-income parents are somehow intellectually deprived and thus in need of an enrichment program, Piaget's theory casts some serious doubt in this area. If the theory is valid—and so far it has received wide support—then it's fair to assume that all children, irrespective of social background, are involved in active experiencing, are organizing and regulating those experiences, and are doing so at comparable levels of complexity. In Piaget's system, intelligence is the level of organization and regulation of actions, not the content that is being regulated. Thus, if there is any way in which we can properly speak of cultural deprivation, it must be done in reference to both the specific contents or actions which a particular culture expects of the child and those with which the child is familiar. The paradox arises that if "culturally deprived" children are put into environments (such as day-care centers) so that they can interact with content that is normally unfamiliar to them, then they will invariably experience an equally debilitating form of deprivation in which it becomes increasingly difficult to function since the normal content and experiences which feed the child's intelligence are no longer present or are no longer valued by the social milieu.

What we need is not a program devoted to compensation, but a program that relates to the knowledge, experience, and processes of the child, and relates in such a way that they can flourish and develop. We need not worry whether they will develop in the direction which the educational community thinks best. If the skills and knowledge that we believe are so important to the child actually have that level of importance, then the child will move towards an acquisition of these skills if he is given the opportunity. But this will take place only if the learning situation responds to the child's already acquired knowledge and capabilities.

FIVE PRINCIPLES OF RESPONSIVE EDUCATION

Over the last four years I have had the opportunity to participate in the development and implementation of a national Head Start and Follow-Through program known as the Responsive Educational Program (16). The term "Responsive" was chosen to reflect our belief that a good educational climate is one where the teacher responds to the learner, rather than the other way around. That is, in "Responsive Education" we are concerned with maintaining a focus on the child's interactions and his intellectual and affective interpretations of those interactions. In my view, this approach is highly consistent with the implications of Piaget's theories on child development and the acquisition of knowledge. What the responsive approach suggests is that the child must play an active interactional role, that he must participate in the elaboration of goals and the development of means of achieving those goals, and that he must play an active role in evaluating the significance and meaning attached to his actions. In short, whatever we wish the child to develop must be achieved through a modification of what he already has, and this can be achieved only if what he has can be exercised in a self-regulatory fashion; that is, in a manner which actually taps the child's mental structures. "Responsiveness" means tapping the child's mental mind as it is and as it suggests it can be in the immediate future and not imposing on the child what it should know at any moment or what it should be able to do in a given time.

In the course of transferring this educational perspective into practice, we have found the writings of O.K. Moore to be particularly useful (17). Moore used a sociological approach to explain the workings of folk models, which he views as the vehicles of noninstitutional education. The fact that culture goes back much further than schools themselves suggests both that culture is somehow learned or transmitted and that it is often accomplished without the direct benefit of institutions such as schools. Moore proposes that folk models account for a great portion of learning and that folk models are essentially forms of play—play at becoming an adult member of the culture.

In the course of developing his sociological model, Moore outlined five principles which describe the character of folk model play. We have adapted these principles to define the conditions for an educational environment. For the present, I would like to discuss Moore's five principles in the context of mental development as I have presented it, leaving for the moment the sociological model of play as outlined by Moore. The five principles are the productive, the perspectives, the responsive, the autotelic, and reflexiveness.

The "productive" principle holds that the knowledge which we present to the child must be organized to enable the child to assimilate it to his intelligence, and not just to

his memory. This means that the information must be organized so that the relationships inherent in the material are
accessible to the child's structures. If this is done, the child
can digest the material in such a way that he comes to see
relationships between its parts and between his assimila-
tion of those parts. If this condition is met, the child's
interactions with the world will generate or produce anticipations and predictions; and, as I indicated earlier, prediction is the direction towards which intelligence develops.
Anticipation is a prediction into the future which is made
possible because the relationships inherent in the past are
understood and suggest the form of the future. When we
guess that the "butler did it" it is because the mystery story
has a structure which we consciously or unconsciously
understand. A pattern that the child can extend is an example of a productive organization if the child can grasp the
relationships inherent in the part of the pattern already
exposed and then figure out what comes next.

The second principle, the "perspectives" principle,
says that each perspective a learner takes, whether active,
passive, evaluative, interactional, or other, yields a different
kind of information about the world and therefore presents
a different reality for the mind to grapple with. In order for
the mind to come to understand something it must be able
to assume a number of perspectives towards it because
each perspective provides a different challenge to the
child's constructive process. A serious criticism is that all
too often education keeps the child in the same perspective, a passive victim of the environment.

The "responsive" principle says that the child must
have a certain amount of control over his interaction with
the environment. He is in the best position to determine
when an activity is appropriate, when he no longer can cope
or focus, when he loses interest, which associations and
relationships to look at, how fast to proceed, what resources to use, and so on. Unquestionably these choices

and freedoms have to take place within a context that provides direction and guidance. The key issue is guidance that helps clarify opportunities rather than dictates activities. The latter is bound to be less efficient in the long run.

A fourth principle is the "autotelic" principle. The word "autotelic" is used to describe situations that are self-motivated—that attract the child's interactions without imposition from others. Thus, one characteristic of autotelic activity is that it is self-rewarding activity. A related aspect of the autotelic principle emphasizes the fact that we can expect the child to bring his intelligence to bear on a problem only if he feels safe in doing so. This safety concerns the emotional or affective risk involved in the situation (the possibility of loss of self-esteem, loss of face, failure and frustration, rejections by others, the appearance of stupidity, etc.) as well as physical risk. This issue is even more important when we are dealing with children from different life styles and cultures. All activities and content carry implicit value statements. If the child is confronted with tasks, materials, or examples that place him in an unfavorable light or comparison, the situation is bound to be less autotelic. The autotelic principle also means that the child should be able to carry out his initial contact with new ideas or tasks in a climate that is free from critical appraisal from the adult world.

The fifth principle is "reflexiveness." It concerns the child's ability to look back upon his actions and to see what he did, how he did it, and so on. It concerns the whole problem of feedback by which the child comes to understand the significance of his actions and how to regulate them. The mind cannot grow without feedback that not only states whether the child is "right" or "wrong" but enables the child to look at his actions after they have been carried out and see more clearly the consequences of his own functioning. This process will always be a component in any area where the child's mental capacities are growing.

We see in these principles, which Moore called "principles for a clarifying environment," the conditions necessary for the support of the development process. As Piaget has suggested, the child grows through the elaboration of goals and means of achieving goals and a great portion of the thrust of this process must be regulated by the growing mind itself. This means that the manner in which the child interacts with the environment, what he chooses to investigate, which perspective he takes, the manner and extent to which he pursues his explorations, are all factors over which he must have some control. If we keep these notions in mind and at the same time create a socially and physically safe environment where the child can receive immediate feedback from his actions, we have come far in supporting mental development. However, operationalizing such an environment at the day-care level or for older age groups is no simple matter. It involves a delicate and intricate balance between freedom and structure; it involves an attitude of tolerance, respect, enthusiasm for, and understanding of children. As we understand more about children it becomes easier to maintain the proper conditions of structure and freedom and to develop attitudes of tolerance, respect, and enthusiasm.

Characteristics of the Day-Care-Age Child: Implications for Education

In furthering this understanding, I'd like to make some broad observations regarding the day-care-age child and the implications of these observations for day-care education, keeping in mind, however, that simple generalizations cannot do credit to the complexity of a four- or five-year development period.

I think one of the important things to keep in mind is that until long after the child has left the day-care setting, play and imitation are the main vehicles for development. Play has long been overlooked as an important means by which the child builds his social and intellectual intelligence. Today, more and more emphasis is being given to understanding the importance of play (3, 18). On the other hand, while imitation has struck many as an important function in development, it has generally been looked at as simply "copying" reality. However, Piaget has shown that this "copy" behavior is a complex constructive process which is essential not only to learning by imitation but to the whole construction of intelligence.

By imitating, the child is forced to grapple with the nature of reality and, as development progresses, we see this imitation becoming more and more refined. In play, the child is freed to distort reality and to apply what he has learned in a manner that is unencumbered by the constraints of reality. Piaget holds that play and imitation go hand in hand; one is not possible without the other. For example, the child who is making motor noises and moving a block of wood around as if it were a car is engaged both in play and imitation. On the one hand, he is imitating the sounds, movements, and functions of an automobile, and on the other, he has distorted the reality of the block of wood and sandpile to become a car and a road. We see in this example that the child's play is a vehicle for experimenting with or trying out his understandings of reality.

Play and imitation—which are hard to distinguish in the early years—are interesting to us because so much learning and adaptation is achieved through their function. At the most basic level, play involves the practice of skills that the child has already acquired. That is, the child plays at what he can do. We find this type of play in animals and even in our adult lives, as when we run through overlearned routines just for the pleasure of exercising our skills. Even this primitive form of play is critical in the child's growing flexibility in the use of his capacities.

Play also serves the child's growing awareness and knowledge about the world. Play is a form of exploration, a means by which the child interacts with the world. And social play, which begins very early with explorations of the mother's body, with peek-a-boo, and so on, constitutes an important source of the socialization process and the child's sense of relationship to actions of others.

Another aspect of socialization involves the means by which the child comes to relinquish his view that the world exists to satisfy his every need, and, in fact, exists by virtue of his needs. Abandoning this view is an important process and one which takes many years to accomplish. And it is also a process that is virtually impossible without the function of play. The child must come to incorporate the prohibitions and needs of others, whether this be using the toilet, not throwing food, or going to bed. The child is eased through this painful process by the use of play. The child invents imaginary playmates to carry out prohibited acts, plays the roles of the parenting adult, and uses play to imitate and express feelings that he is trying to learn how to control. In play, the child's feelings and needs are reduced to a symbolic and imaginary form, and hence removed from the child's direct and immediate struggle.

Language is another area of development that is strongly affected by play. In fact, Piaget has suggested that language and play constitute the same process. Language is generally viewed as structured system or signifiers and signifieds—that is, words, gestures, and so on that stand for or represent, and the objects, actions, and so on that are represented. In a sense, words stand for reality; in play, the child's actions stand for reality. As in language, where words are used in more accurate ways, so, too, is play moved closer to an accurate representation of reality. Beyond the common properties shared by language and play, the two are brought together in process as well. From their earliest occurrence, words are used in play, and it is in this context as much as any other that the meaning and power of words are explored.

Even in this brief presentation, we see that the function of play is important to growth. I am not suggesting that all learning is play, or that it is even fun; or that by play alone, all learning will take place. But it remains that play is the most powerful vehicle for learning which the young child has, and that the conditions which support play may well deserve our attention in considering the kinds of conditions we'd like to see in early childhood educational environments.

Returning to Moore's five principles, we can see how, in the process of play, the child himself supplies all the conditions for responsive education. Since play is being structured by the child, the productive condition is automatically met and we see the consequence of this in the child's anticipations or in his "I know what I'm doing and where I'm going" attitude. The taking of different perspectives is always involved in play. Children act upon the environment, take passive roles, become critics and references, and seldom are found to be involved in their play from the same perspective. As to the autotelic principle, play by its very nature is autotelic, that is, self-motivated, and carried out in a make-believe world that minimizes psychological risks. The responsive condition is also met when the child plays at whatever comes to mind or appeals to fancy, moves the play at his own place, quits when he wants to, and so on; could we conceive of play in a situation where every factor is dictated? Last, the reflexiveness principle. The spontaneous nature of play suggests that the child is constantly integrating feedback into directing and channeling the play. So we see that play is very important to the child's learning. And of course, play should be viewed as a legitimate and integral part of the day-care curriculum. In fact, I would suggest that almost all of the "cognitive goals" of day care could be achieved through the use of play. But it is also important to note that, just as with intelligence, play is also a developmental phenomenon. That is, it serves functions and goes through stages of development in which it becomes increasingly complex both in its forms and functions. Again, as with intelligence, this growth is, on the whole, a self-regulated phenomenon. That does not mean that the development of play does not require stimulation from others—including adults—but it does mean that certain environmental conditions need to exist in order for this development to occur. Moore's principles for a clarifying environment provide a good description of these conditions for the development of play as well as intelligence itself.

Another issue of importance to day care concerns the child's acquisition and use of language. As indicated earlier, true use of language as a system of reference (symbols that stand for things) does not occur until the end of the sensory-motor period, at which time the child evidences many expressions of a representative capacity. While language will be used from this time onward, how it is used changes with age. Piaget characterizes the first few years of the child's language use as being "egocentric" (10). Put as briefly as possible, the child spends more time talking to himself than to others. That is, until somewhere between three and four years of age, language is used minimally as a form of social interaction. For example, when children of this age are playing together, you may find that they engage in parallel conversation—neither party talking to the other, but both carrying on conversation in their mutual presence. During this period the child is using language to help represent, and thereby regulate, his own actions and feelings, not to coordinate these functions with the activities of others. This does not mean that the child's use of language is not social. He encodes language to express to others and decodes to interpret the intentions of others. But these two functions are not yet coordinated, and the child's predominant language activity is "talking to himself."

Around four years of age, the child has mastered most of the grammatical regularities of language. Up to this point the child's syntax is limited and primitive. By four, the child has learned to regulate language to form such elements as possessives, various tenses, and plurals. With this achievement, but not necessarily because of it, the child begins to use language in a truly social manner—to engage in a genuine sharing of viewpoints. However, even at this late stage the use of language still lacks a systematic and logical character. Returning to the earlier example about the use of the term "brother," we see that for the young child, the word "brother" is not the same as it as for the adult—a concept of a set of parallel relationships in which to have a brother is to be a brother. In any area where we find a lack of consistent and "logical" thought, we also find a lack of consistent and logical use of language. This prelogical thought lasts until the child is seven or eight years of age. Even at this time, the language system is still undergoing significant modification which will not be completed until late adolescence.

The significance for day care is twofold. First, we should realize that language is also a developmental phenomenon and that it grows by virtue of its use. Therefore, the day-care setting should provide verbal and linguistic stimulation. However, and second, we should realize that the language is not fully formed during this period, that it is used for different functions at the early stages, and that it is not capable of conveying or interpreting thought of which the child himself is still incapable. That is, there is a close relationship between language and thought such that the use of language, irrespective of its apparent adult form, does not mean that the child's understanding is not still that of a child.

While I've talked about the lack of logical thought in the young child, this should not be interpreted in a limited fashion. The child's world is a bizarre place in which all manner of phenomena are possible. The child sees human-like forces and motives behind diverse events—the movements of objects by the wind, the presence of shadows, the working of machines—and sees intelligent life in inanimate objects like rocks and mountains (11). The child's conception of space leads to real concerns over going down the bathtub drain and other such "silly childhood fears." I recently came across a story of a three year old who looked forward to flying to the east coast with his parents and seemed to have a clear picture of this upcoming event. However, as the departure approached, the child became depressed and one day, in a deluge of tears, blurted out his fears. He had not yet "learned to fly" and could not go to "Urp" with his parents. To be responsive to the child at this age, adults must realize that the world does not appear to the child as it does to the adult. We must constantly "listen" with all our facilities if we are to be capable of responding to the child's reality as he sees and experiences it.

Another area of great concern to those associated with young children involves the socialization process—the child's movement towards civilization and "appropriate behavior." It should be understood that from the child's perspective, this is a painful process of giving up freedom and an omnipotent view that the world runs to his tune. We realize that for the benefit of society and the child, this view must be relinquished. But the child is not persuaded by our arguments. Arguments and rationality do not civilize children. What does is the establishment of clear prohibitions and limits, presented in degrees which the child can follow and understand and in ways that offer a reason for doing so. However, the reasons are not rational but affective or emotional ones. The most powerful force behind the internalization of other's desires and limits is the child's desire to maintain the love of those who are most important to him. The child takes on the desires of others through a desire to please and be accepted by others. When the internalization has occurred, the limits no longer are observed simply to please others but to please the child himself. What someone else wanted now becomes something the child himself wants, and the motive behind this involves affective issues of trust, acceptance, and love. The same process can be accomplished through punishment and the child's desire to avoid pain, but the consequences are generally repressed behaviors which are bound to show up elsewhere, whether in bedwetting or social withdrawal.

In all the unstable and untrustworthy aspects of the world, the one thing the child counts on most is stability and trust in his relations with significant adults. To maintain that relationship is vital for the child and the price he pays, even when it appears at the surface to be unwillingly, is socialization, the march towards civilization. This is one of the challenges to programs for children: that is, providing these relationships within the child-care context.

NOTES

1. Piaget, J. and Inhelder, B. *The Psychology of the Child.* New York: Basic Books, 1969.
2. Piaget, J. *The Construction of Reality in the Child.* New York: Basic Books, 1954.
3. _____. *Play, Dreams, and Imitation in Children.* New York: Norton, 1962.
4. _____. *The Origins of Intelligence in Children.* New York: Norton, 1963.
5. _____. *The Psychology of Intelligence.* Totowa, N.J.: Littlefield, Adams, 1963.
6. _____. *Structuralism.* New York: Harper & Row, 1970.
7. _____. *The Child's Conception of Numbers.* New York: Norton, 1965.

8. _____, and Inhelder, B. *The Child's Conception of Space.* New York: Norton, 1967.
9. Inhelder, B. and Piaget, J. *The Early Growth of Logic in the Child: Classification and Seriation.* New York: Harper & Row, 1964.
10. Piaget, J. *The Language and Thought of the Child.* New York: Meridian Books, 1955.
11. _____. *The Child's Conception of Physical Causality.* Totowa, N.J.: Littlefield, Adams, 1965.
12. _____. *The Child's Conception of the World.* Totowa, N.J.: Littlefield, Adams, 1965.
13. _____. *The Moral Judgment of the Child.* New York: Free Press, 1965.
14. _____. *Judgment and Reasoning in the Child.* Totowa, N.J.: Littlefield, Adams, 1966.
15. _____, and Inhelder, B. *Mental Imagery in the Child.* New York: Basic Books, 1971.
16. Nimnicht, G. P. "Overview of the Responsive Program." In *Beyond 'Compensatory Education': A New Approach to Educating Children,* ed. by G. P. Nimnicht and J. A. Johnson, Jr. Washington, U.S. Government Printing Office, 1973. Pp. 110–125.
17. Moore, O. K. and Anderson, A. R. "Some Principles for the Design of Clarifying Educational Environments." In *Handbook of Socialization Theory and Research,* ed. by David A. Goslin. New York: Rand McNally, 1969. Chapter 10.
18. Almy, M., editor. *Early Childhood Play.* New York: Selected Academic Readings, 1968.

Chapter Thirteen

PROVIDING FOOD SERVICES

Linda Regele-Sinclair

> So closely and intricately interwoven into a strand within the
> individual are both physical nature, which requires food and
> that nature which we call intellectual development, that it
> will not do to keep them separate. (1)

The children of any society are its wealth, for in them is
embodied the hope of a culture for a better future. If children are denied access to good, nutritious food they also
will be denied an opportunity to become fully involved in
their environment. Therefore, the first concern of any day-
care program should be to ensure that each child receives
the proper amount of food needed to sustain both physical
and intellectual growth.

Because children in day care receive a substantial part
of their daily food needs in the center, much attention

should be given to the planning and operating of a meal program.

It is equally important for day-care centers to encourage children to become interested in food and actively involved in its preparation. It may be said then that food not only sustains life but it also can educate a child about life.

The relationship between human growth and proper nutrition has long been understood. A study done by John Boyd-Orr in 1936 demonstrated that the height of a controlled group of children varied according to the social class of the child. He suggested that the difference in height could be attributable in part to heredity but that "environmental differences particularly related to nutrition were probably highly influential in producing the observed gradient in height." (2)

Other investigations, such as those showing the relationship between inadequate nutrition in early life and its effect upon the growth of tissues and organs have added to an ever-growing body of knowledge surrounding proper nutrition and its relationship to growth development.

The foods that children eat will affect not only their physical growth but also their own sense of well-being, their joy in being alive, and their ability to learn. A child can discover much about himself through food. Foods are symbolic and can carry a strong emotional message for children. Probably most importantly, foods represent love and security.

These are important concepts for homes or centers which care for children during the day to keep in mind when planning a meal program. For as Evans, Shub, and Weinstein state in their book *Day Care,* "Any day care facility has the responsibility to provide a program which facilitates children's optimal growth and development." (3)

But just how does a day-care home or center ensure an interesting and adequate food program?

WHERE TO BEGIN: OR HOW TO GET FOOD TO YOUR CENTER

For most centers the prospect of running an adequate meals program means finding more money. More money may mean increasing fees charged to parents or investigating what is available to centers in federal food assistance.

The Special Food Service Program for Children, sponsored by the Department of Agriculture (USDA), is designed to help nonprofit private and public day-care and Head Start centers operate meal programs. It provides cash, commodities, and equipment money to any eligible, nonprofit center for the operation of a food program.

The Special Food Service Program (Section 13 of the National School Lunch and Child Nutrition Act) was introduced in Congress by Representative Charles Vanik (Ohio) in early 1968. Later that year the bill became law. It authorized a pilot food service program for day care to be operated for three years.

With the passage of the Special Food Service Program for Children legislation, Congress and USDA took their first step towards seeing that children cared for outside of their homes would be guaranteed the right to nutritious food and good health.

The Special Food Service Program is administered by the USDA. The USDA has the responsibility to formulate the rules and regulations which define how the program is to operate. It also administers the program on the state level. The state department of education is responsible for collecting claims, reimbursing centers, and providing technical assistance wherever it is needed. Sixteen state departments of education are prohibited by law from administering the programs outside of the public school system. In each of those states, the regional office of the USDA assumes the state agency's responsibility.

Originally Congress intended that the program should focus on centers serving primarily children from low-income families. Today the program has been broadened and the Special Food Service Program is available to:

> any private, nonprofit institutions or public institutions such as child day care centers, settlement houses or recreation centers which provide day care, or other child care where the children are not maintained in residence for children from areas in which poor economic conditions exist and from areas where there is a high concentration of working mothers. (4)

The program includes private, nonprofit institutions providing day-care services for handicapped children. Unfortunately, family day-care homes are not presently eligible for the Special Food Service Program.

The Special Food Service Program for Children provides a cash reimbursement for each meal served plus varying amounts of donated commodities. If a state so chooses, it may set aside up to 25 per cent of its apportioned funds for food preparation and storage equipment to be used by participating centers or potential program sponsors. This nonfood assistance is made available in the form of one-time grants for the purchase of kitchen equipment or, in the case of leased equipment, as a regular subsidy for a given period of time. Many states have chosen not to allocate money for equipment and have thereby restricted program expansion.

Some day-care and Head Start centers may also qualify as "especially needy" centers. Centers showing severe need must demonstrate that the per meal reimbursement rate received is insufficient to carry on an effective feeding program. Centers designated especially needy can receive "financial assistance not to exceed 80 per cent of the operating costs of such a program, including the cost of obtaining, preparing and serving the food."

To apply for participation in the Special Food Service Program, a center must fill out a series of applications and forms. Upon approval the center must sign a statement agreeing to abide by the rules and regulations for basic operation of a meals program. Under the regulations children must be guaranteed the right to receive a free or reduced price meal, if they are eligible. Centers may not discriminate against children receiving free and reduced price meals and must also agree to prepare meals which meet the specified meal requirements set by the USDA. These are referred to as Type A meal standards.

Type A meals provide for minimal necessary nutritional intake. The department suggests that centers use the following as meal guidelines:

> *Age: 1–3, breakfast*—1/2 cp. milk, 1/4 cp. fruit or juice, 1/2 slice bread or equivalent or 1/4 cp. cereal
>
> *lunch/supper*—1/2 cp. milk, 1 ounce meat or equivalent quantity of an alternative, 1/4 cp. vegetable, fruit or both, 1/2 slice bread or equivalent
>
> *snacks*—1/2 cp. milk or juice, 1/2 slice bread or equivalent
>
> *Age: 3–6, breakfast*—3/4 cp. milk, 1/2 cp. juice or fruit, 1/2 slice bread, 1/2 cp. cereal or equivalent
>
> *lunch/supper*—3/4 cp. milk, 1 1/2 ounces meat or equivalent, 1/2 cp. vegetable, fruit or both, 1/2 slice bread or equivalent
>
> *snacks*—1/2 cp. milk or juice, 1/2 slice bread or equivalent (nuts, fruits, etc.).

After a center has been approved for assistance it will receive a reimbursement of 36¢ for lunch or supper; 18¢ for breakfast; and 12¢ for snacks. If commodities are available, centers will receive what the state delivers.

The federal government also sponsors the Special Milk Program, which is open to any nonprofit private or public center. Applications for the Special Milk Program are available from each state department of education. The program currently provides centers with 5¢ for every pint of milk served. Centers may also be reimbursed for the full cost of the milk if it is served to a child who is eligible for free meals.

The responsibility of a center for providing a meal program does not end with getting the financing for food and seeing that it is prepared. A center should build a meals program which embodies all the benefits of food.

FOOD, MEALS, AND CHILDREN'S NEEDS

Bruno Bettleheim once said, "Food is the greatest socializer." Food can serve as an excellent vehicle for learning. Involving a child in preparing and eating a meal offers the opportunity of enhancing and enlarging his world.

Children learn about the world by doing, feeling, seeing, and tasting. Food can serve as a good example of texture, shapes, and color. It can help to teach word skills and new concepts. Buying and preparing food helps the child learn math skills. Scientific concepts of life and growth can be taught by planting a vegetable or herb garden. A sense of love, sharing, care, and communication can be expressed and experienced in the actual eating of the meal.

> Food consumption is strongly influenced by custom habit. Each cultural group tends to have its own strong preference for certain foods and cooking. People eat the food they like, when they can get them; that is, they eat the food which they find attractive in taste and flavor. . . . Each cultural group tends to regard its own habits as the normal and natural ones . . . food habits are, in fact, very deeply rooted in each culture. (5)

Children's tastes may be broadened by introducing them to a wide range of foods. The preparation of ethnic foods is a particularly good way not only to broaden children's tastes but to help them learn about other people and cultures. A day-care center in Minneapolis that has Native American, Afro-American, and white children has published a small recipe book with the favorite recipes of each child who participates in the center. Each parent was asked for his or her child's favorite recipe (one which was simple enough for the children to help prepare). Each day a different meal was prepared, stressing whose favorite food it was, what culture it came from, and so on. This experience served both as a means of communication and an introduction to different foods for all the children in the center.

In any meals program a center should see that the children enjoy eating their food and are getting nutritionally adequate meals. The best way to do this is to consider the children's food needs.

The Maryland state licensing agency suggests that "a variety of foods and snacks which appeal to children and which meet children's daily nutritional needs shall be served at intervals of not more than three weeks apart." (6)

The meal should be an enrichment of the child's total personality. If the mechanics of eating are difficult, a child may become frustrated and lose any interest or desire to eat. Therefore, tables, chairs, and utensils should be manageable, as should the size of the food portion. If a child requires a particular diet, all measures should be taken to ensure that the child gets the necessary meal and at the same times does not feel left out.

The involvement of teachers, aides, and staff in preparing and eating the meal is critical. A meal shared by all can be a highly rewarding experience. It can offer everyone —children and adults alike—a chance to interact and communicate with one another around a common, shared experience.

It is also important for the staff to closely examine how the meal experience is used. Are meals a time of socializing and educating or are meals used as bribes, punishments, and rewards? Do staff members eat with the children—and do they eat the same foods as the children?

Snacks can also be a time during which staff and children interact. A snack might be a simple food that the children help to prepare. The preparation experience can encourage new knowledge about foods, their preparation, texture, and odor. Children also learn skills in measuring, in mixing, and in following directions.

Food and its importance in a day-care program should not be confined to just meals and snacks. Other activities such as field trips related to food; stories, games, and songs about food; discussions and participation in food preparation; parties, holidays, and ethnic foods; food-related art work; and growing vegetables can help to further extend children's interest in and knowledge of foods.

How Can a Center Continue to Make Food an Interesting and Special Part of the Day?

To focus on food as one of the more important educational tools, the staff of a center may feel that continual information is needed—and it is. Not everyone is born a nutritionist or an economist or a sociologist or knows what kind of food assistance is available to centers and how to apply for it.

This is an area where, potentially, the state licensing office can play an important, progressive role in providing centers with sources of information which will ensure quality care. Most states and counties have nutritionists on their public health staff available on a consulting basis. Many county agent offices have home economists on staff also available upon request. Use them.

A small number of the day-care center staff, with the assistance of the resource people, could put together a series of workshops, which could be offered both to day-care center staff and parents. The workshops could include nutrition information as it relates to the food supply, production, and distribution; the politics of food; food assistance programs; control of the food economy and its affect on food costs; and eating habits.

There are many resources, both public and private, which can be tapped to assist in putting together and running food workshops. If a center or centers are interested in pursuing nutrition/food education on a more long-term basis, there are various funding sources which could be tapped to support these efforts.

Locally, centers could contact the United Way, local foundations, the Jaycees, Kiwanis Club, the American Legion, and local businesses to request nutrition education money. Prior to contacting any potential funding source, the center(s) should prepare a brief statement of their problem, a plan of how the funding source can help correct the problem, and a cost figure.

Very often funding sources make money available on a one-time basis. Some thought should be given to sources of continual funding if food and nutrition training is held to be a useful endeavor.

A WORD ABOUT FAMILY DAY-CARE HOMES

Although this article has focused primarily on day-care centers, children and nutrition should be no less important a concern in family day-care homes. Family day-care mothers and fathers should take the same time and care in planning meals and food activities as do day-care center staff. In some ways the day-care mother or father is at a distinct advantage in having the opportunity to more fully involve the children in preparing and eating the meal.

Yet family day-care homes are not eligible to participate in the Special Food Service Program. Therefore, family day-care mothers and fathers are required to rely upon the fees received from parents and money from their own pockets to pay for food. Now there is a great deal of interest in seeing that family day-care homes are included under new SFSPC legislation.

Regarding nutrition and food-training activities planning, the situation varies from county to county, city to city, state to state. Some areas send extension agents into the homes of family day-care mothers and do on-the-spot training. In other areas there are no organized services available to family day-care mothers. And in still other areas, family day care is well organized and mothers attend classes, receive certificates in early childhood education, and continually participate in workshop updates.

Although training varies from place to place, attempts should be made both by the family day-care mother and responsible agencies to see that children in family day-care homes are accorded the same food and nutrition benefits as children in day-care centers.

Food is the basis of life. It is one of the better socializers for children. Without good food and adequate nutrition starting at an early age, children are handicapped in their ability to learn and experience the world around them.

Yet offering adequate food to children is not enough. Children are constantly tempted by fast foods. Formulated foods are rapidly replacing good, wholesome, natural food. Advertising preys on children's minds, enticing them to demand sugar-coated cerals and oversweet cakes or soft drinks. And always, our society says, *fast*—eat fast, eat on the run.

Children must be, as we all must be, educated to make good food choices. A day-care center or family day-care home which realizes this need and does something about it can offer much more to the growth of a child than two pieces of enriched white bread with bologna could ever do.

Notes

1. Bettleheim, Bruno. *Food to Nurture the Mind.* Washington: The Children's Foundation, 1970. p. 23.
2. Birch, Herbert G., "Malnutrition and Early Development." Edith Grotberg, editor, *Day Care: Resources for Decisions.* Washington, D.C.: Day Care and Child Development Council of America.
3. Evans, E. Belle, Shub, Beth, and Weinstein, Marlene. *Day Care: How to Plan, Develop, and Operate a Day Care Center.* Boston: Beacon Press, 1971. p. 113.
4. 1973 Regulation for Special Food Service Program for Children, p. 4, 225.7(b).
5. Aykroyd, W. R. *Food for Man,* London: Pergamon Press, 1964. p. 75.
6. Maryland State Licensing Regulations, "Food Service."

Bibliography

Amidon, Edna A. *Good Food and Nutrition.* New York: John Wiley, 1946.

Aykroyd, W. R. *Food for Man.* London: Pergamon, 1964.

Barry, Erick. *Eating and Cooking Around the World.* New York: John Day, 1963.

Bettleheim, Bruno, *Food to Nurture the Mind.* Washington: The Children's Foundation, 1970.

Birch, Herbert G. "Malnutrition and Early Development." *Day Care: Resources for Decisions,* ed. Edith Grotenberg. Washington: Day Care and Child Development Council of America.

Borgheses, Anita. *The Down to Earth Cookbook.* New York: Scribner's, 1973.

Child Welfare League of America. *Standards for Day Care Service.* New York: Child Welfare League of America, 1968.

Croft, Karen B. *The Good for Me Cookbook.* San Francisco: R. & E Associates.

Evans, E. Belle, Shub, Beth and Weinstein, Marlene. *Day Care: How to Plan, Develop, and Operate a Day Care Center.* Boston: Beacon Press, 1971.

Fenton, Carol and Kitchen, Nermine. *Plants that Feed Us.* New York: John Day, 1971.

Goodwin, Mary and Pollen, Gerry. *Creative Food Experiences for Children.* Washington: Center for Science in the Public Interest, 1974.

Hille, Helen H. *Food Groups for Young Children Cared for During the Day.* Washington: Children's Bureau Publication #386, Supt. of Documents, USGPO, 1960.

Levine, Lois. *The Kids in the Kitchen.* New York: Macmillan, 1968.

Maryland State Licensing Regulations, "Food Service," Baltimore.

Maternal and Child Health/Maryland State Department of Health *Newsletter.* Baltimore: March, 1965.

Tannahill, Reay. *Food in History.* New York: Stein & Day, 1973.

Chapter Fourteen

THE ORGANIZATION OF DAY CARE: CONSIDERATIONS RELATING TO THE MENTAL HEALTH OF CHILD AND FAMILY

**Christoph M. Heinick,
David Friedman,
Elizabeth Prescott,
Conchita Puncel,
and June Solnit Sale**

President Nixon's veto on December 9, 1971, of the Comprehensive Child Development Act with its day care provisions has further focused a previously existing issue: How shall the extension of these day care services be carried

*The American Orthopsychiatric Association Study Group on Mental Health Aspects of Day Care

**Presented at the 1972 annual meeting of the American Orthopsychiatric Association, in Detroit.

†The work of the authors was supported in part by Office of Child Development Grant Nos. 48 (Heinicke), R-219 (Prescott), and CB-10, Continuation-2 (Sale), and by the Foundation for Research in Psychoanalysis, Los Angeles (Heinicke).

out? We are concerned that in the ensuing political debate the exchange of slogans will interfere with realistic and badly needed solutions. We hope that the statements made below will clarify what can optimally be realized through the expansion of child care services.

President Nixon focused his concern on a number of areas. Three of the most important run parallel to the three broad topics covered in this paper:

1. He asked what type of program, if any, as designed by various child related professionals is truly desirable for the child? To what type of child development program should "the vast moral authority of the National Government be committed?" We are also concerned with the extremes of unstimulating custodial care versus highly structured curricula designed to control and modify only one part of the child's behavior, *e.g.*, the cognitive. In the first section, we discuss developmental criteria for evaluating a day care program. We trust this discussion will lay to rest charges that a developmental approach involves "brain washing" and is preoccupied with "detecting difficulties" in the child.

2. The President in his veto message was troubled by the extent to which the resources and rights of home and family might be ignored in favor of an extensive and expensive bureaucracy. We are also concerned with the extreme views on day care that imply that parents are incompetent and that any form of entrance into the home is better than none, or, on the other hand, completely ignore the family's resources and fail to give that type of support that will enhance a family's self-respecting initiative. We hope our discussion will reveal that the developmental approach outlined here strengthens the parent's role rather than "invading the home" or shifting responsibility from parent to federal government.

3. The President also asked how we can organize extensive care so that the needs of children and families, rather than those of an expensive bureaucracy, remain the priority. We do not approve of packaged delivery systems that are "efficient" but do not in fact meet individual community needs. And we are concerned that the child care giver replacing the parent during the day have the kind of administrative support that will allow him or her to function adequately with the child.

Although day care is most frequently discussed in relation to the group day care center, we here include all other arrangements made to take care of the child during the whole or part of the day. We shall stress not only the importance of diversity of services but also the possibility that new and various combinations of arrangements will be devised.

We foresee that many experienced professionals will disagree with our position statements. The difference of viewpoints will be due to differences in value judgments, general experience with children and families, and certain research findings. Hopefully, the ensuing dialogue will generate further exploration and research, and will contribute to the national debate that President Nixon called for in his veto message.

DEVELOPMENTAL CRITERIA FOR EVALUATING A DAY CARE PROGRAM

Developmental in the context of mental health will here refer to the growth of the child's total functioning. Because of its great importance, the developmental point of view outlined here will also require consideration of those environmental and, especially, family influences that make an impact on the child. Any day care service must be evaluated in terms of how it enhances the external and internal factors that impinge on the child and lead to a greater number of progressive as opposed to regressive developments. Does the day care program assess and *promote* the child's growth? Does the day care experience *enhance* the child's positive family experiences? Does the day care program offer important *supplemental* or *alternate* experiences to those available in the family environment?

  1. *That all aspects of the child's development be considered in planning and evaluating a day care service.* It is central to our point of view that all aspects of the child's functioning—physical, nutritional, social, motivational, and intellectual—be considered in formulating a day care service. This point of view has most recently been eloquently developed by Zigler.[37] There is much evidence[3,10,32] to indicate that cognitive development is intertwined with social and emotional development.

  In turning away from the idea of custodial care, many have ardently promoted structured curricula in order to insure skills needed for school entrance. In this approach the instructor decides what is to be learned or changed and makes most of the choices as to the conditions under which learning will take place. Quite apart from the question of promoting choice-making, there is no consensus that skills are more effectively taught by structured approaches.[33] Using IQ gains and reading achievement as measured in third grade, Weikart[35] has recently reported that the nature of the curriculum, cognitively oriented, language oriented (Bereiter-Englemann), or unit based (child centered-traditional) was not a significant variable. If anything, the structured approach of Bereiter-Englemann was less effective than the other two in skill training as reflected in reading scores.

  Nor can the assessment criteria be confined to such measures as IQ or reading. The child centered unit approach may well have an advantage in the social and motivational areas of development. *Planned* but less structured approaches, if that planning incorporates what is of interest to a particular child, are likely to lead to multiple gains. Thus, the reading of a book chosen by the children, if followed by questions and discussion with the children, not only enhances concept learning, but is likely to develop curiosity, listening skills, and a feeling of pride if the questions are answered correctly. That is, we are very much questioning the long run efficiency of working with very selected response systems and are assuming that in a world of accelerating possibilities the ability to make choices will be critical.

2. *That the experiences planned for the child be maximally suited to his individual developmental needs.* To follow an individualized approach, whether it be in developing language skills or enhancing the child's creative expression in relation to art materials, it is important that his strengths and weaknesses in all essential areas of development first be established. Clearly, there are practical limits to what a given staff can do in such assessment and individualizing, but it must remain as a goal rather than giving in to efficient packaging for large groups. Children may learn in unison certain words or skills but when the teacher and/or group is absent this growth disappears.[16] Children may conform in the face of powerful routines designed for handling large day care groups, but one must ask what is their inner control when no longer part of such group pressure.[24] It is critical to the child's developing interest in the world and his individual style of dealing with it, that individual adult attention be available to encourage, guide, and challenge him.[26]

A further assumption in developing the individualized curriculum is that one begins with and expands those areas of interest where the young child is already involved. These starting points can be used to move in the direction of further building the child's strength and removing certain areas of weakness. Specifically, this means that at certain points one gives the child a choice of where to be involved, and then follows him through whatever he has chosen. If he consistently manipulated playdough, this could be expanded to deal with concepts of weight, the pleasure of texture, and as an opportunity for generalizing this engagement to engagement in other activities and relationships. It also means that in a group situation at juice time, interest in a graham cracker could become the focus of a discussion of size or experience of taste. Or one can build on the natural interest of the child in "who is present and who is absent" to develop concepts of counting or to communicate and respond to the child's concern about his immediate separation from parents and other family members or to relate the function of adult caretakers in the day care setting to the child's ongoing life in his own home. Or the first sense of achievement and self may come in the area of climbing the junglegym, and can then be expanded to interest in other areas. Also, one certainly helps the child to begin to learn in such structured situations as listening to a book being read and answering questions about it. When, however, learning is confined to certain lesson plans, too much of the time is often spent reprimanding the child to listen.[17,27]

3. *That the day care program encourage the child to explore, make choices, and develop a variety of coping methods.* Following White,[36] we assume that the child's tendency to explore and master is central in the development of competence and must be given maximum opportunity. Inevitably this will lead to some conflict between children and between children and adults. Yet the exploration of the total environment and the successful resolution of conflict leads to the development of coping methods that can be used to adapt to new situations. For example, when two children are fighting over who had the wooden spoon first, it may

be tempting to resolve the situation by removing the spoon altogether. Entering into the situation and discussing it may accentuate the angry feelings, but it also teaches modes of resolution.

The opportunity to explore on one's own, engage in one's own pursuit, and form one's own social groups even if at times conflictual must also be viewed as likely to promote that inner feeling of: "This is me; this is who I am; this is what I did." It is important that the program indeed facilitate experiences that lead to a clear and positive self-image.

4. That the day care program promote the development of a variety and balanced expression of feelings—joy, anger, pride, sorrow, affection, sympathy, etc. Confronted with large numbers of young children whose impulse life is still labile, a day care teacher may understandably move a program towards greater structure and defined time schedules. There is, however, considerable evidence now to indicate that neither the excessively authoritarian nor the permissive child-rearing environment is likely to promote maximum development.[2] Although not enough is as yet known about what kinds of experiences do lead to the child's balanced expression of feelings, its importance can be stressed not only from an experiential or value point of view, but also because it has been found to be central to a cluster of personality dimensions predictive of the continuing adaptation of the child. Named "ego flexibility", it includes the balanced availability of a variety of affects, the balanced use of defenses, the absence or repression of aggression, and nondefensive use of humor.[10]

Returning to the evaluation of day care programs, a program that permits the excessive expression of feeling is not likely to promote the balanced experience and expression of that feeling. On the other hand, a program that puts a premium on compliance, order, and quiet also is not likely to promote that goal. Prescott and Jones[26] have shown that children expressed less enthusiasm, wonder, and pleasure in day care centers where teacher restraint and control was high.

The implication of the statements above can be made more explicit by examining two functions that our experience as well as the research literature indicate can serve as important criteria for evaluating whether a day care program does enhance the sound development and thus the mental health of the child. The first function is the child's task orientation; the second is his ability to move psychologically from his previous environment and become involved in the new day care environment.

5. *That the total day care experience enhance the child's task orientation.* Task orientation refers to the manner in which the child engages, produces, and takes pride and pleasure in a task. It is concerned with commitment and the results of that commitment. Tasks here include self-defined tasks such as sequential purposive play in a doll corner, the self-chosen task of completing an art task with materials provided by the child or teacher, adult tasks flexibly defined, such as making a collage with materials provided by the teacher, and more strictly defined tasks like answering a question about a story read by the teacher to a group of children. In assessing the child's task orientation, the following subcategories have been developed: the engagement in the task; compliance with the definition of the task situation; the quantity and quality of productivity; signs of creativity and imagination in pursuing the task; and pride and pleasure in the results.[11]

We have focused on the concept of task orientation for several reasons. It first of all reflects well our concern with motivational and attentional processes as well as those considerations relating to productivity. Second, it has been demonstrated that this task orientation can be assessed fairly readily in something like the listening-to-a-book time. Third, it has been shown[1,19,21] that variations in the child's task orientation, or components of it, are highly predictive of later learning, specifically the ability to read. More recent unpublished work[14] shows that the nature of the task orientation at about three-and-a-half is related to other indices of adaptation to nursery school and is in turn significantly associated with the indices of task orientation at four and five years of age and a minimal discrepance between performance and verbal IQ as measured by the WPPSI at five years.

If the above findings clearly need further substantiation, they do encourage us to consider the child's ability to become task oriented as an important focus in evaluating his total development. However, far less is known about the type of day care program that enhances the child's task orientation. Intensive observation and filming of two girls entering a day care program[13] does suggest the hypothesis that the availability of a continuing one-to-one relationship with a teacher's aide, in addition to the teachers and other personnel, may well be a critical factor in developing the kind of commitment to learning central to an adequate task orientation.

6. That the total day care experience facilitate the psychological move from the previous family caretaking environment to the new day care environment. Even if the previous caretaking environment is minimally supportive it is important to consider how the child will move from adaptation to it and become successfully involved in the new day care environment.[34] Research on the three-year-old entering a preschool setting[14] indicates that the child who can show his longing for his parents but yet sustain the expectation that he will be cared for and can then make new commitments to the adults and peers in the nursery school tends to be the child who will show the best total development, including the development of an adequate task orientation. The same research approach applied to three-year-olds entering day care suggests:

1. The manner in which mother and child enter the day care program reflects an essential component of their relationship.
2. The nature of that relationship, and especially whether the child has the inner expectation that he will be cared for, is very significant in determining whether he can make use of the new relationships offered in the day care program.
3. The continued availability at the center of a caretaker, such as a grandmother to whom the child has a relationship, will greatly facilitate the transition from care in the home to care in the day care center.
4. The impact of the loss of parent is likely to be less if the child has a chance to adjust to the half-day care program before being moved to the full-day care stay.[9]

In summary, our developmental criteria for the evaluation of day care programs stress attention to *individual needs,* attention to *all areas of functioning* (and not just, for example, the cognitive one), promotion of the child's *active choice* in what he does and learns, encouragement in learning to deal with a *variety of feelings,* and enhancement of the quality of *engagement* as opposed to passive receiving. We believe neither a program designed to take care of large numbers of children and preventing flexible planning nor the highly structured program that does not allow the child to define his interests is as likely to meet these criteria.

In evaluating a given approach to day care, it is important to ask whether or not there has been an adequate assessment of the previous experiences of the child and whether the total program has provided continuity by sustaining the previous experience in building, supplementing, and providing new ones.

INTERRELATION OF THE FAMILY AND THE TYPE OF DAY CARE PROGRAM

This brings us to the question of how best to integrate the strengths of existing family and child care resources into the day care program. It is again assumed that success in this regard will be most beneficial to the mental health of all involved.

   7. That day care services supplement and complement the family rather than replacing that basic caretaking unit. Beginning with the model of the extended family, and noting that many family units no longer have the support of relatives, it becomes a matter of positive social policy and innovation to create services that complement, supplement, and, in this sense, share the task of child-rearing. This includes preventive health and mental health measures and definitive health and mental health services, especially if health and mental health professionals participate in the planning of the program and in staff orientation. It is neither economically feasible nor psychologically wise to set up model programs that communicate that "we can do it better" and, in effect, exclude the existing family unit. The day care experience that does not enhance the development of family and neighborhood is likely to be very limited, if not confusing. One need not be romantic about the Western family structure to state that current information suggests that in various forms it continues in many instances to provide those intimate supportive and stimulating experiences that are conducive to sound child development.[15]
   It follows from the above stress on the extended family model that the director, teachers, and other staff have an intimate and continuing relationship to both family and child. Moreover, as these relationships develop, it will be necessary that the members of this "extended family" share their experience in order to instill a sense of working to-

gether. This is one specific meaning of the concept of consumer control. For example, if the teacher's aide has developed the kind of closeness that allows the child to talk to her about his innermost thoughts just before naptime, it is important that a way be found of sharing the essence of this information with the mother or father.

8. *That, in a given community, a family have the true option of using family day care or group care or some other combination of arrangements.* If we take the model of the extended family and individualized planning seriously, then the conception of a day care system based only on group care must be questioned. We propose that a diversity of options is a more useful approach. This is another meaning of the concept of consumer control. Our experience and research [22,29] tell us that, for many families, family day care is an optimal arrangement. We believe this to be particularly true of the child up to the age of three. In addition, family day care can accommodate the needs of families with a variety of ages. Small, home care settings do accept children with minor ailments (such as a cold), which assures parents stability in their work situations and makes it more likely for the parent to report the ailment, so that measures such as rest and quiet may be provided the child. Lack of public transportation and long distances to group day care centers often make a neighborhood family day care facility, located next door or in the next block, more desirable. In the small groupings of a family setting, a child is able to feel himself more personally responsible and responded to (positively and negatively). In addition, cultural and ethnic value systems between users and givers of a service in a neighborhood are more likely to match. The encouragement of a strong system of good child-rearing environments in a neighborhood setting serves the function of "horizontal diffusion" described by Klaus and Gray.[19]

Traditionally, group care has only served healthy children from ages two to five. Recently, high quality group centers offering infant care have been developed by such workers as Caldwell,[3] Provence,[28] and Keister;[18] many other demonstration programs dealing with various ages are now under way. Ways will have to be found of duplicating such centers at a cheaper rate without loss of quality. Hopefully, these demonstration projects will suggest new options for families with different needs. Various combinations of group, family day care, and in-home care should be tried.

Like group centers, family day care also varies in quality and is sometimes woefully inadequate. Demonstration efforts are again under way to improve the quality by providing supporting services and enhancing the self-esteem and economic stability of the family day care mother.[6,31]

9. *That continuing professional help be available to the family and be defined by both the professional and the family members.* There are times when intervention by day care staff is necessary to assist the child's development. The most extreme case is child abuse. The staff should also include someone trained to cope with minor emergencies and to observe the child for health aberrations. There may be other instances where specialized services (*e.g.,* physician or dentist) must be called in. It may be possible for a health professional who is well integrated into the total program to introduce concepts of prevention and anticipatory concepts in the areas of health and nutrition. However, in areas of behavior, especially those relevant to central family values, it is important to recognize that there are limitations to what mothers and fathers will accept from "members of the extended family." Thus, a teacher may feel a boy should be more assertive and competitive, but this may conflict with attitudes that the family has about aggressiveness, war, toy guns, etc.

The first general point to stress in this regard is that many families do want help but they also want to participate in the definition and delivery of that help. The preschool programs successfully administered by Weikart[35] and his colleagues all included home visits during which the professionals assisted with problems that the families were trying to solve.

Closely related, a second general point then is that the help to the family be given by someone who is adequately trained to do that job. While different situations will require different forms of training, we feel it essential that there be an awareness of the meaning of giving help, the nature of the relationship involved, and the limitations imposed by considerations of the need for privacy and confidentiality. We are not speaking of training for psychotherapy, but have found that supportive work with the parents must be guided by an understanding of the inner and external situation of the person being helped. A third general point is that the help be offered from the time the family enters day care and that it be understood that a contact will be maintained on a regular basis.

10. That the basic child-rearing pattern associated with certain ethnic and racial groupings be fully recognized in organizing a particular day care service. It is consistent with our emphasis on day care as a supplement and complement of family care that one be particularly responsive to the needs of the particular subgroupings being served. These needs are often not expressed, or are perhaps even put in ways that are misleading. It is essential that the person working with a particular group know and accept differences in these needs and in ways of dealing with them. Similarly, the day care director should be sensitive to the implications of the type of care she develops for the self-esteem of the family she is serving. For example, are they constantly being asked about their financial incapacity? Or, if the much-needed research is undertaken, is care taken that the people involved benefit from the research and related services? Most important, the subgroup differences should first of all be accepted as assets rather than indicating a deficit. Having to exist in two different cultures can result in an identity crisis for the child. A day care center that truly communicates the worth of the child's family and heritage is most likely to assist in the resolution of such conflicting feelings.

11. That in planning help for the family, one give special attention to family-child relationship experiences found to be critical to sound child development. As already indicated, to implement meaningful assistance to families, it is, of course, imperative to assess what will be useful and what will be acceptable. The family's wish to determine what help they want, and the great racial and ethnic differences that exist, must be respected. But one should also be guided in one's understanding and planning by the accumulating knowledge in regard to factors influencing sound child development. For example, a variety of research does suggest that the competent functioning of a child in preschool is associated with a certain profile of parent-child relationships. The preschool children who showed an adequate separation from their families, a commitment to new relationships, and also developed an adequate task orientation, tended to come from families with a mother who could provide warmth, whose psychological availability was very clearly defined, who encouraged new relationships, who definitely set limits, and who, in addition, exerted pressure for independent achievement, particularly in the pre-academic area.[14] These findings are very consistent with those of other studies[2] that demonstrate links between child care practices and competent functioning of the preschool child. Rutter[30] has shown that active discord and a lack of affection in the family are associated with antisocial disorder, but a good relationship with one parent can mitigate the effect of a quarrelsome home.

12. That the day care program include those experiences not available in the home and felt to be critical to the development of the children being cared for. While stressing the priority of developing family resources, if it is determined that these are inadequate and cannot be improved, the day care program may have to be designed to supplement the child's experiences in certain areas.

For example, if warm acceptance by the mother and clear availability is important in something like the development of task orientation, and if this is difficult to enhance in what the parents give, then it may indeed be critical to provide this in the day care setting. For some other children, the problems resulting from the absence of an adult male in a single parent home may make it important to provide experiences with a male in the day care setting.

If one focuses on the process whereby the child increasingly explores the world as well as his own ability to cope with it and to define himself in relation to it, then it is also important to consider whether at moment of anxiety there is a "homebase" to which he can return physically and psychologically. It is particularly the affectionate, holding stable kind of care that is likely to be insufficient in the large, group day care center, and more available in the family day care arrangement. It can, however, also be provided in the group center. We believe that this component of the child's experience is indeed central. There is much evidence[2,15] to indicate that the adequate provision of this single component is essential to the healthy development of the child.

Another example of designing the day care service to supplement family resources would be to provide the child with the nutrients missing in his diet, while at the same time working with the family to encourage healthy food habits.

13. That a structure be developed that insures adequate day-to-day relevant communication between child care staff and parent. If the total experience of the child is to be one that builds modes of coping with the expectable and understandable, then some effort must be made to help the main caretakers communicate with each other. Thus, the parent must know that a child has perhaps that day been struggling with his angry feelings toward a peer. Similarly, the day care staff must be made aware by the parents of the critical events occurring in the home. The information exchange need not be of a lengthy sort, but there should be some expectation that such information will be exchanged.

14. That one member of the caretaking staff be fully informed about the total development of the child and family, and be the advocate of necessary changes to promote that development. It is central to the point of view being developed here, namely the attention to all aspects of the functioning of the child, that someone also insure the total and meaningful integration of experiences making an impact on the child. This idea is consistent with the principle of advocacy that has been so central to the report of the Joint Commission on the Mental Health of Children.

It follows from the above that a day care program provide comprehensive rather than single services to a child and family. It is also consistent with these considerations that the day care program include some efforts to provide information on resources for families not immediately involved in the program. Volunteers have been effectively used to provide such information.

In summary, using the model of the extended family, we have stressed the importance of the day care services *complementing,* not replacing, the family resources. Even though expensive, we have recommended that an adequately trained professional attempt, first of all and on a continuing basis, to help the families define and solve their own problems. In those areas where this is too difficult, attention may well have to be focused on the daily work with the child in the day care setting; for example, providing the warmth and limits not available in the home.

ADMINISTRATIVE SUPPORT FOR PROGRAM CONTENT

In this section focus is placed on the day to day program content and its relation to the administrative values and practices governing that program. The effectiveness of a program cannot be judged simply by its descriptive content but must be related to the administrative policy that governs the day care service. Elaborate curriculum plans issued by agencies that are far removed from local needs may do little to enhance the development of the children if the administrative policies are, in fact, not congruent with these plans. Weikart[35] has stressed the importance of regular and encouraging supervision in the successful carrying out of his preschool programs.

These principles of support apply to family day care as well. Collins and Watson[4] have suggested that the quality of family day care is improved by providing support through social work consultation for a centrally placed "day care neighbor." Sale[31] has demonstrated that quality may be improved by providing support to an existing network of family day care homes through a variety of direct and community based resources. In addition, a self-help organization of family day care mothers might prove to be a method of improving services for children, their families, and the providers of day care.

15. That administrative practices support autonomy in decision-making on the part of staff responsible for children. Children need adult models who can make decisions, accept failures, and innovate resourcefully. Administrative systems designed in the interest of safety, efficiency, and control often remove the responsibility for decision-making from the level at which the decisions must be carried out. As Cummings and Cummings[5] have demonstrated, such removal tends to introduce ambiguity and ambivalence into relationships at all levels.

16. That staff working with children receive stimulation and support from their administration. Working with young children for long periods of time with only limited adult contact can be a boring and tiresome job. If caregivers are to be expected to individualize their care and to support task orientation through their relationship with each child, careful thought needs to be given to the supportive environment that is provided to them. Opportunities to work with outside educational, health, and mental health consultants, to take trips, and, most of all, to use one's own ideas in building programs, all help the caregiver to see her job as exciting and important.

The provision of a workable physical environment can also be a most important form of support. Prescott[27] has found that play areas that have good organization, variety, and sufficient amounts of play equipment of adequate complexity are associated with large amounts of teacher encouragement. In play areas that were rated low on these criteria, teachers spent much more time in restriction and guidance, and children's responses were generally low. Teachers also need adequate storage space, and convenient access to play areas that are easy to maintain. Activities that offer possibilities for real construction projects, such as a carpentry table with *real* tools, ramps, and large wooden boxes that permit building and high climbing, often are particularly important to boys. Animals and messy activities such as wet sand, dirt, fingerpaint, and clay provide other kinds of opportunities for children but are more work for teachers and often will not be made available unless administrators give consideration to their importance and provide the settings and support so that they can be offered.

The physical setting also needs to provide opportunities for real conversation about questions that may never reach formulation without careful nurture. In interviews with day care teachers, Prescott[25] found that they seldom reported answering questions about birth, procreation, racial differences, fear, etc. Teachers often need administrative help and support to tackle emotion-laden questions and to figure out ways of picking up on children's puzzlement. Equally important is the provision of an environment that provides a wide range of life experiences and the provision of cozy, private, intimate settings in which meaningful conversation can occur.

17. That restriction of activity because of possible safety hazards be carefully considered in terms of developmental goals. Children who have opportunities to explore at their own speed, in environments peopled by helpful adults, learn to handle increasingly complex activities with safety and dexterity. In the interests of safety, many day care centers have chosen to eliminate swings, climbing apparatus, tools such as saws and hammers, and innovative play such as riding tricycles backwards. Such concerns for safety tend to reduce the complexity of the environment and limit opportunity for development of responsibility and physical competence. If the setting provides insufficient challenge, children often are goaded into limit-testing that actually can be dangerous.

18. That the size of the day care center be kept in the 30–60 range in order to enhance it as a child-rearing environment. While there are clear financial and administrative advantages to organizing a large day care center (100 or more children), research by Prescott[24] has shown that the quality of the child-rearing environment decreases in the following ways: In large, as compared to small, centers, there was less flexibility in scheduling, teachers devoted more effort to control and restraint, and teachers were more often rated as distant or neutral and less often rated as sensitive to the needs of the child. In large centers, there were fewer opportunities for children to initiate and choose activities; richness of the play was thus curbed. Most telling in the large, as opposed to small, center was the lack of interest and enthusiasm found among children. That is, those aspects of what is here termed task orientation and, more broadly, commitment are more prevalent in the children of the small center.

In summary, the administrative structure and delivery

system must be guided primarily by the individual needs of
the children and the proper balance of support and autonomy for day care givers, rather than by the efficiency criteria of providing as many slots per dollar as possible.
Experience suggests that this is most likely to occur in the
smaller group center and in the family day care setting with
adequate support for the day care giver.

Concluding Remarks

Because of its profound implications, the extensive provision of day care should indeed be the subject of national
debate. To consider adequately the real needs of our nation's young children, we must focus on the contemporary
scene as it exists, and not be detracted by charges of "brain
washing" and "sovietization of our children." On the basis
of our research, we have recommended:

1. Criteria for the evaluation of day care programs,
    which stress:
    a. Attention to individual needs.
    b. Attention to all areas of functioning.
    c. Promotion of the child's active choice-mak-
        ing.
    d. Encouragement of learning to deal with a
        variety of feelings.
    e. Enhancement of engagement as opposed to
        passive receiving.
2. A model for interrelating family and day care
    based on the concept of the extended family and
    implemented by:
    a. *Extensive communication* focusing on the re-
        sponsiveness of the day care givers to the
        needs of the family, and encouraging the par-
        ticipation of families.

b. *Support of family resources* by both the day care giver and continuing, well trained professional help.

c. *Supplementing of family-to-child experiences* in the day care setting, especially in critical areas where direct help to the family is not feasible.

3. A delivery system not guided by the concept of the most slots per dollar, but focusing on the children's needs and in particular on the proper balance of support and autonomy for the day care giver. This is presently likely to be true in small group day care centers and adequately supported family day care. We wish to stress, however, that different models of child care need to be explored and evaluated. Various combinations of group, family day care, and in-house care may well take advantage of existing resources and best meet family needs.

References

1. Attwell, A., Orpet, R. and Meyers, E. 1967. Kindergarten behavior ratings as a predictor of academic achievement. J. School Psychol. 6:43–46.
2. Baumrind, D. 1970. Socialization and instrumental competence in young children. Young Children 26:104–119.
3. Caldwell, B. et al. 1970. Infant day care and attachment. Amer. J. Orthopsychiat. 40:397–412.
4. Collins, A. and Watson, E. 1969. The Day Care Neighbor Service: A Handbook for the Organization and Operation of a New Approach to Family Day Care. Tri-County Community Council, Portland, Oregon.
5. Cummings, J. and Cummings, E. 1967. Ego and Milieu. Atherton, New York.
6. Emlen, A. 1970. Neighborhood and family day care as a child-rearing environment. Paper presented at the annual meeting of the National Association for the Education of Young Children, Boston.

7. Friedman, D. et al. 1959. Water, Sand and Mud as Play Materials. National Association for the Education of Young Children, New York.
8. Gordon, I. 1969. Stimulation via parent education. Children 16:57–59.
9. Heinicke, C. 1971. Parental deprivation in early childhood: a predisposition to later depression? *In* Separation and Depression: Clinical and Research Aspects, E. Senay, ed. Symposium to be published by the American Association for the Advancement of Science.
10. Heinicke, C. 1969. Frequency of psychotherapeutic session as a factor affecting outcome: analysis of clinical ratings and test results. J. Abnorm. Psychol. 74:553–560.
11. Heinicke, C. et al. 1971. Parent-child relations, adaptation to nursery school and the child's task orientation: a contrast in the development of two girls. *In* Individual Differences in Children, J. Westman, ed. In press.
12. Heinicke, C. et al. 1971. A methodology for the intensive observation of the preschool child. *In* Individual Differences in Children, J. Westman, ed. In press.
13. Heinicke, C. et al. 1972. Relationship opportunities in day care and the child's task orientation: three-year-old Amy and Julie. Unpublished manuscript.
14. Heinicke, C., Liebowitz, J. and Aronson, L. 1972. Adaptation to nursery school, the child's task orientation, and performance on the WPPSI. Unpublished manuscript.
15. Hess, R. 1969. Parental behavior and children's school achievement: implications for Head Start. *In* Critical Issues in Research Related to Disadvantaged Children, E. Grotberg, ed. Educational Testing Service, Princeton, N.J.
16. Karnes, M., Teska, J. and Hodgins, A. 1968. A longitudinal study of disadvantaged children who participated in three different preschool programs. ERIC Document Reproduction Service, ED 036–338.
17. Katz, L. 1969. An approach to language learning. Young Children 24:368–379.
18. Keister, M. 1970. "The Good Life" for Infants and Toddlers. National Association for the Education of Young Children, Washington, D.C.
19. Klaus, R. and Gray, S. 1968. The early training project for disadvantaged children: a report after five years. Monogr. Soc. Res. Child Develpm. 33(4).
20. Levenstein, P. 1970. Cognitive growth in preschoolers through verbal interaction with mothers. Amer. J. Orthopsychiat. 40:426–432.

21. Lindemann, E. et al. 1967. Predicting school adjustment before entry. J. School Psychol. 6:24–42.
22. Low, S. and Spindler, P. 1968. Child care arrangements of working mothers in the United States. Children's Bureau Publication No. 461–1968. Superintendent of Documents, U.S. Government Printing Office, Washington, D.C.
23. Peters, A. 1971. Health support in day care. *In* Day Care Resources for Decisions, E. Grotberg, ed. Office of Economic Opportunity, Washington, D.C.
24. Prescott, E. 1970. The large day care center as a child-rearing environment. Voice for Children. 3(4).
25. Prescott, E. 1964. Children in Group Day Care: The Effect of a Dual Child-Rearing Environment. Research Department, Welfare Planning Council, Los Angeles Region.
26. Prescott, E. and Jones, E. 1971. Day care for children: assets and liabilities. Children 18:54–58.
27. Prescott, E. and Jones, E. 1967. Group day care as a child-rearing environment. USDHEW, Pacific Oaks College, Pasadena, Calif.
28. Provence, S. 1969. Children under three . . . finding ways to stimulate development. Children 16:53–55.
29. Ruderman, F. 1968. Child Care and Working Mothers: A Study Made for Day-Time Care of Children. Child Welfare League of America, New York.
30. Rutter, M. 1971. Parent-child separation: psychological effects on the children. J. Child Psychol. Psychiat. 12:233–260.
31. Sale, J. 1971. I'm Not Just a Babysitter. Pacific Oaks College, Pasadena, Calif.
32. Skeels, H. 1966. Adult status of children with contrasting early life experiences: a follow-up study. Monogr. Soc. Res. Child Develpm. 31(3).
33. Spiker, H. 1971. Intellectual development through early childhood education. Exceptional Children 37:629–640.
34. Van Leeuwen, K. and Pomer, S. 1969. The separation-adaptation response to temporary object loss. J. Amer. Acad. Child Psychiat. 8:711–733.
35. Weikart, D. 1971. Early Childhood Special Education for Intellectually Subnormal and/or Culturally Different Children. National Leadership Institute in Early Childhood Development, Washington, D.C.
36. White, R. 1963. Ego and reality in psychoanalytic theory: a proposal regarding independent ego energies. Psychological Issues, Monograph II, 3(3).
37. Zigler, E. 1970. The environmental mystique: training in the intellect versus development of the child. Childhood Ed. 46:402–412.

APPENDIX

A. Additional References
B. Model Centers and Child-Care Systems
C. Additional Resources
D. Guide to Resources for Volunteers
E. Health Component Improvement Plan Key

APPENDIX

Appendix A: Additional References on Model Programs and Planning

American Joint Distribution Committee. *Joint Distribution Committee Guide for Day Care Centers: A Handbook to Aid Communities in Developing Day Care Center Programs for Pre-School Children.* Geneva: 1962. (ERIC Publication No. EDO 27961.) [ERIC Clearinghouse on Early Childhood Education, University of Illinois at Urbana-Champaign, 805 West Pennsylvania Avenue, Urbana, IL 61801.]

Atkinson, Jonathan. *Day Care Costs in Massachusetts.* Boston, MA: State of Massachusetts, March, 1973. 15 pp. [Office for Children, ATTN: Massachusetts State 4-C Committee, 120 Boylston Street, Room #246, Boston, MA 02116.]

Auerbach, Stevanne and Freedman, Linda. Choosing Child Care—A Guide for Parents. San Francisco, CA. 94103 1976. 80 pp. $3.00 (Parents and Child Care Resources, 1855 Folsom St., San Francisco, CA. 94103)

Bedger, Jean E.; Ehrlich, Lawrence D.; Zemont, Delia; Silhavy, Carol Kelpsas; and Weed, Catherine. *Financial Reporting and Cost Analysis Manual for Day Care Centers, Head Start, and Other Programs.* Chicago, IL: Council for Community Services in Metropolitan Chicago.

March 1973. 185 pp.: $7.50. [64 East Jackson Boulevard, Chicago, IL 60604.]

Bernstein, Blanche and Giacchino, Priscilla. "Costs of Day Care: Implications for Public Policy," *City Almanac* 6:2. New York: New School for Social Research, August, 1971.

Bikle, J. "Social Casework in a Day Care Program," *Day Care: An Expanding Resource for Children*. New York: Child Welfare League of America, Inc., 1965. Pp. 36–43. [67 Irving Place, New York, NY 10003.]

Caldwell, Bettye M. "What Is the Optimal Learning Environment for the Young Child?" *American Journal of Orthopsychiatry* 37 (January, 1967): pp. 8–22.

Center for the Study of Public Policy. *An Impact Study of Day Care: Feasibility Report and Manual for Community Planners*. Cambridge, MA: February, 1971. 198 pp. [56 Boylston Street, Cambridge, MA 02138.]

The Child Care Task Force Appointed by the City Council. *A Plan for Child Care in Palo Alto*. Palo Alto, CA: City of Palo Alto, Division of Reproduction, April, 1973. 83 pp.

Child Welfare League of America, Inc. *Child Welfare League of America Standards for Day Care Service*. New York: 1969. 123 pp.: $2.50. [67 Irving Place, New York, NY 10003.]

Committee on Infant and Preschool Child. *Recommendations for Day Care Centers for Infants and Children*. Evanston, IL: American Academy of Pediatrics, 1973. [P.O. Box 1037, Evanston, IL 60204.]

Cohen, Donald J. *Day Care: 3: Serving Preschool Children*. Washington, DC: U.S. Government Printing Office, 1974. 164 pp. [Superintendent of Documents, U.S. Government Printing Office, Washington, DC 20402.]

Early Childhood Task Force. *Early Childhood Planning in the States: A Handbook for Gathering Data and Assessing Needs*. Denver, CO: Education Commission of the States, 1973. 46 pp.: $1.00. [300 Lincoln Tower, 1860 Lincoln Street, Denver, CO 80203.]

_____. *Establishing a State Office of Early Childhood Development: Suggested Legislative Alternatives*. Denver, CO: Education Commission of the States, 1973. 48 pp.: $1.00. [300 Lincoln Tower, 1860 Lincoln Street, Denver, CO 80203.]

Educational Resources Information Center (ERIC). *Day Care: An Annotated Bibliography*. Urbana, IL: June, 1971. [ERIC Clearinghouse on Early Childhood Education, University of Illinois at Urbana-Champaign, 805 West Pennsylvania Avenue, Urbana, IL 61801.]

_____. *Directory of Resources on Early Childhood Education*. Washington, DC: Day Care and Child Development Council of America, Inc. 1971. (Reprint.) [1401 K Street NW, Washington, DC 20060.]

Emlen, Arthur C.; Donoghue, Betty A.; and LaForge, Rolfe. *Child Care by Kith: A Study of the Family Day Care Relationships of Working Mothers and Neighborhood Caregivers.* Portland, OR: Portland State University, 1971. 332 pp. [Field Study of the Neighborhood Family Day Care System, 2856 Northwest Savier, Portland, OR 97210.]

Farson, Richard. *Birthrights: A Bill of Rights for Children.* New York: Macmillan Publishing Co., Inc., 1974. 248 pp.: $6.95. [866 Third Avenue, New York, NY 10022.]

Food Research and Action Center. *Out to Lunch: A Study of USDA's Day-Care and Summer Feeding Programs.* Yonkers, NY: Gazette Press, Inc., 1974. 94 pp.: $2.00.

Food and Nutrition Service. *A Guide for Planning: Food Service in Child Care Centers.* Washington, DC: U.S. Department of Agriculture, 1971. 22 pp.: $.55. [Superintendent of Documents, U.S. Government Printing Office, Washington, DC 20402.]

Goldsmith, C. "A Blueprint for a Comprehensive Community-Wide Day Care Program," *Child Welfare* 44 (1965): 501–503, 528.

Grosett, Marjorie D.; Simon, Alvin C.; and Stewart, Nancie B. *So You're Going to Run a Day Care Service!* New York: Day Care Council of New York, Inc., Autumn, 1971. 88 pp. [114 East 32nd Street, New York, NY 10016.]

Harrell, James A., ed. *Selected Readings in the Issues of Day Care.* Washington, DC: Day Care and Child Development Council of America, Inc., 1972. 85 pp. [1401 K Street NW, Washington, DC 20060.]

Hest, M. S. "A Broad Community Approach to Day Care," *Child Welfare.* 39 (1960): 29–32.

Hoffman, Lois Wladis and Nye, F. Ivan. *Working Mothers: An Evaluative Review of the Consequences for Wife, Husband, and Child.* San Francisco, CA: Jossey-Bass, Inc., 1974. 272 pp.: $12.50. [615 Montgomery Street, San Francisco, CA 94111.]

Idaho Office of Child Development. *Growing Up In Idaho: The Needs of Young Children.* Idaho Office of Child Development. Boise, ID: State of Idaho. 17 pp.

Kiester, Dorothy J. *Consultation in Day Care.* Chapel Hill, NC: Institute of Government, University of North Carolina at Chapel Hill, 1969. 72 pp.

Kitano, Harry H. L. *The Child-Care Center: A Study of the Interaction Among One-Parent Children, Parents, and School.* Berkeley and Los Angeles, CA: University of California Press, 1963. 344 pp.: $1.25.

Knight, E. V. *Serving the Pre-School Child: Day Care as a Service to the Entire Family.* New York: National Federation of Settlements and Neighborhood Centers, 1966. [232 Madison Avenue, New York, NY 10016.]

Kraus, Dorothy; Beal, Elaine; Taylor, Rose; and Blackford, Lillian. *Survey of Pre-School and Day Care Needs and Facilities in San Mateo County.* San Mateo County, CA: Office of San Mateo County Superintendent of Schools, San Mateo Board of Education, 1970. 50 pp.

League of Women Voters. *Day Care: Who Needs It.* Washington, DC: 1973.

Lewis, L. "Broad Community Approach to Day Care," *Child Welfare,* 39 (1960): 32–33.

Massachusetts Early Education Project. *Child Care in Massachusetts: The Public Responsibility.* Washington, DC: Day Care and Child Development Council of America, Inc., 1972. [1401 K Street NW, Washington, DC 20060.]

Mattick, Ilse and Perkins, Frances J. *Guidelines for Observation and Assessment: An Approach to Evaluating the Learning Environment of a Day Care Center.* Washington, DC: Day Care and Child Development Council of America, Inc., January, 1973. 44 pp. [1401 K Street NW, Washington, DC 20060.]

Morgan, Gwen G. *Regulation of Early Childhood Programs.* Washington, DC: Day Care and Child Development Council of America, Inc., January, 1973. 130 pp. [1401 K Street NW, Washington, DC 20060.]

Moss, Mary Ann; Sullivan, Rhita Jean; and Pratt, Eleanor. *A Menu Planning Guide for Type A School Lunches.* Washington, DC: U.S. Department of Agriculture, 1974. 20 pp. and foldout worksheet: $1.00. [Superintendent of Documents, U.S. Government Printing Office, Washington, DC 20402.]

Murphy, Lois B. and Leeper, Ethel M. *Caring for Children—No. 6: A Setting for Growth.* Washington, DC: U.S. Department of Health, Education, and Welfare, 1973. 24 pp.: $.55. (Illustrated.) [Superintendent of Documents, U.S. Government Printing Office, Washington, DC 20402.]

_____. *Caring for Children—No. 7: The Individual Child.* Washington, DC: U.S. Department of Health, Education, and Welfare, 1973. 24 pp.: $.55. (Illustrated.) [Superintendent of Documents, U.S. Government Printing Office, Washington, DC 20402.]

Office of Child Development. *State and Local Day Care Licensing Requirements.* Washington, DC: U.S. Department of Health, Education, and Welfare, August, 1971. 52 pp. and Appendices A-J: $1.75 domestic postpaid; $1.50 GPO Bookstore. [Superintendent of Documents, U.S. Government Printing Office, Washington, DC 20402.]

Prescott, Elizabeth. "A Comparison of Three Types of Day Care and Nursery School-Home Care." Paper presented at the Biennial Meeting of Society for Research in Child Development, Philadelphia, PA. March 20—April 1, 1973. 11 pp. [Elizabeth Prescott, Pacific Oaks College, 714 West California Boulevard, Pasadena, CA 91107.]

Ruderman, Florence A. "Day Care: A Challenge to Social Work," *Child Welfare*. 43 (1963): 117–123.

Ruffino, Barbara C. and Aitken, Sherrie S. *Day Care in Maryland: A Study of Child Development Needs and Resources* (Summary Report). Washington, DC: Maryland Child Development Planning Project, Maryland Office of Child Development, March, 1972. 5 pp. [733 - 15th Street NW, Washington, DC 20005.]

Ruopp, Richard R. *A Study in Child Care, 1970–71.* Cambridge, MA: ABT Associates, Inc. Washington, DC: Office of Economic Opportunity, 1971. (Volume I - *Findings;* Volume II-A - *Center Case Studies;* Volume II-B - *System Case Studies;* Volume III - *Cost and Quality Issues for Operators.*) [ABT Associates, Inc., 55 Wheeler Street, Cambridge, MA 02138.]

Sauer, Peter H. and Hickey, M. F. *Building a Day Care Center: An Introduction to Planning and Financing a Day Care Center: What to Look for and How to Buy It or Rent It with a City Lease.* New York: Bank Street Day Care Consultation Service, February, 1970. [Bank Street College of Education, Day Care Consultation Service, 610 West 112th Street, New York, NY 10025.]

Schultze, Charles L.; Fried, Edward R.; Rivlin, Alice M.; and Teeters, Nancy H. *Setting National Priorities: The 1973 Budget.* Washington, DC: The Brookings Institution, 1972. 468 pp. [1775 Massachusetts Avenue, Washington, DC 20036.]

Senate Committee on Finance (Russell B. Long, Chairman), 2nd Session of the 93rd Congress. *Child Care: Data and Materials.* Washington, DC: U.S. Government Printing Office, October, 1974. 258 pp.: $2.55. [Superintendent of Documents, U.S. Government Printing Office, Washington, DC 20402.]

Shannon, William. "A Radical, Direct, Simple, Utopian Alternative to Day-Care Centers." *The New York Times Magazine.* April 30, 1972.

Siedman, Eileen. *Day Care in Vermont: An Evaluation of the Vermont Model FAP Child Care Service System.* Washington, DC: Leadership Institute for Community Development, 1972. 440 pp.

Swenson, Janet P. *Alternatives in Quality Child Care: A Guide for Thinking and Planning.* Washington, DC: Day Care and Child Development Council of America, Inc., 1972. 79 pp. [1401 K Street NW, Washington, DC 20060.]

Texas Department of Community Affairs. *Early Childhood Development in Texas: 1973–74.* Austin, TX: December, 1973. 159 pp. [Jeannette Watson, Director, Office of Early Childhood Development, Texas

Department of Community Affairs, P. O. Box 13166, Capitol Station, Austin, TX 78711.]
United States Department of Health, Education, and Welfare. *Project Head Start: Nutrition-Staff Training Programs.* Washington, DC: 1969. 36 pp. [Superintendent of Documents, U.S. Government Printing Office, Washington, DC 20402.]
Zamoff, Richard B. *Guide to the Assessment of Day Care Services and Needs at the Community Level.* Washington, DC: The Urban Institute, July, 1971. 100 pp.: $3.00. [Publications Office, The Urban Institute, 2100 M Street NW, Washington, DC 20037.]

APPENDIX B: ADDITIONAL MODEL PROGRAMS: CENTERS AND SYSTEMS

In 1970 ABT Associates, Inc. of Cambridge, Massachusetts received a sizable contract to evaluate child-care centers and systems throughout the country. The report included 13 centers and 7 systems which were described in great detail. For your information these programs are included here.

Centers

Berkeley, CA
Children's Centers
Early Childhood Education
2031 Sixth Street
Berkeley, CA 94710

Charlotte, NC
Child Development Day Care Centers
427 West Fourth Street
Charlotte, NC 28202

Frankfort, KY
Kentucky Child Welfare Research Foundation
314 West Main Street
Frankfort, KY 40601

New York, NY
Family Day Care Career Program
349 Broadway
New York, NY 10013

Pasco, WA
Northwest Rural Opportunities
110 North Second Street
Pasco, WA 99301

Springfield, MA
    Springfield Day Nursery
    103 Williams Street
    Springfield, MA 01106

Systems

Casper, WY
    Casper Day Care Center
    804 South Wolcott Street
    Casper, WY 82601

Chicago, IL
    Amalgamated Day Care Center
    323 South Ashland
    Chicago, IL 60607

Chicago, IL
    Fifth City Preschool
    Ecumenical Institute
    3444 West Congress Parkway
    Chicago, IL 60624

Dorchester, MA
    AVCO Day Care Center
    188 Geneva Avenue
    Dorchester, MA

Greeley, CO
    Greeley Parent-Child Center
    925 B Street
    Greeley, CO 80631

Nashville, TN
    American Child Centers, Inc.
    Woodmont Center

2001 Woodmont Boulevard
Nashville, TN 37215

New York, NY
West 80th Street Day Care
458 Columbus Avenue
New York, NY 10024

Salt Lake City, UT
Central City Head Start Day Care Center
615 South 3rd Street East
Salt Lake City, UT 84111

San Francisco, CA
Haight-Ashbury Children's Center
1101 Masonic Avenue
San Francisco, CA 94117

Syracuse, NY
Syracuse University Children's Center
100 Walnut Street
Syracuse, NY 13210

Washington, DC
Georgetown University Hospital Day Care Center
3800 Reservoir Road NW
Washington, DC 20007

Appendix C: Additional Organizations

American Academy of Pediatrics
P.O. Box 1037
Evanston, IL 60204

American Home Economics Association
2010 Massachusetts Avenue NW
Washington, DC 20016

American Nurses Association
2420 Pershing Road
Kansas City, MO 64108

1030 - 15th Street NW
Washington, DC 20005

American Psychological Association
1947 Rosemary Hills Drive
Silver Spring, MD 20910

Association for Childhood Education International
3615 Wisconsin Avenue NW
Washington, DC 20016

Black Child Development Institute
1028 Connecticut Avenue NW
Washington, DC 20036

Child Development Associate Consortium
7315 Wisconsin Avenue, #601
Washington, DC 20014

Child Welfare League of America
67 Irving Place
New York, NY 10003

Children's Defense Fund
1763 R Street NW
Washington, DC 20009

The Children's Foundation
1028 Connecticut Avenue NW, #614
Washington, DC 20036

Council on Social Work Education
345 East 46th Street
New York, NY

Day Care and Child Development Council of America, Inc.
1401 K Street NW
Washington, DC 20060

National Association for the Education of Young Children
1834 Connecticut Avenue NW
Washington, DC 20009

National Association of Social Workers
20 E Street NW
Washington, DC 20001

National Council of Organizations for Children and Youth
1910 K Street NW
Washington, DC 20006

National Federation of Settlements and Neighborhood
Centers
232 Madison Avenue
New York, NY 10016

National League for Nursing
10 Columbus Circle
New York, NY

APPENDIX D: A BRIEF GUIDE TO RESOURCES FOR VOLUNTEERS Prepared by James A. Levine

Clearinghouses for Volunteers

Volunteers in Technical Assistance (VITA), 3706 Rhode Island Avenue, Mount Rainier, Maryland 20822. Phone: 301-277-7000. Contact: Ms. Lois Schoenbrun.

With funding from the Grant Foundation, VITA has initiated a matchmaking service to link programs and volunteers with expertise in all areas of program planning and operation: curriculum, staff training, facilities design, fund-raising, etc. Programs are referred to volunteers in their geographical area. VITA pays volunteer expenses related to technical assistance, including telephone, travel, and up to $25 per day for food and lodging.

National Center for Voluntary Action, 1625 Massachusetts Avenue, N.W., Washington, D.C. Phone: 202-797-7800.

The national center can refer you to its local affilitate, usually a Voluntary Action Center. The VAC's serve as clearinghouses to match program needs and volunteer skills. Both the national center and the local chapters have staff and guidebooks to provide advice on recruiting and working with volunteers.

Accounting and Business Skills

American Institute of Certified Public Accountants (AICPA), 666 Fifth Avenue, New York, N.Y. 10019. Phone: 212-581-8440. Contact: Manager, Accounting Aid Program.

National Association of Accountants, 919 Third Avenue, New York, N.Y. 10022. Phone: 212-371-9124. Contact: Manager, Socioeconomic Programs.

The above organizations or their state chapters will help you locate one of the many accountants willing to donate services. However grateful you may be for such help, don't expect or let your accountant begin work unless you've taken the time to explain your program and to convey its "feeling." As important as sound business practices are to day care, day care is not the same type of business to which your accountant may be accustomed.

Architects

Your local chapter of the American Institute of Architects (1735 New York Avenue, N.W., Washington, D.C. 20006. Phone: 202-785-7300) may have a willing volunteer. Depending on the level of skill or certification required to meet your needs, you may be able to enlist the services of an architecture student.

Carpentry and Mechanical Work

Local Trade School or Vocational High School. Usually well supervised, the students from these schools take great pride in their work. They are often in demand and well booked in advance.

Architecture Department of a Technical Institute, College, or University. Some of these students can get academic credit for helping you.

College Fraternities. When and where they still exist, some have substituted community-service projects for hazing of pledges. The key to their helpfulness is leadership; unless they have a strong and skilled leader riding herd, the brothers may not show up when they've promised.

ROTC and National Guard. Like the fraternities, the armed services are going through some changes. They are sometimes available for community-service projects.

The previous sources are handy when parents, staff, board members, and friends are overburdened and you need help building shelves or constructing a playground.

Health Screening Services

Your local health department should be able to refer you to free resources. Commonly available are: dental screening, arranged by the women's auxiliary of local dental societies; hearing testing, arranged by societies for speech and hearing or societies for the deaf; vision screening, provided by Lion's Clubs across the nation.

Legal Services

The only organizations working specifically to help daycare programs are legal services and legal aid. If you have difficulty obtaining a legal services lawyer, contact Ms. Marcia Cleveland, Day-Care Coordinator, New York City Legal Services, 335 Broadway, New York, N.Y. 10013 (212-966-6600). In addition, your local bar association may be able to refer you to lawyers willing to do pro bono work.

Nutrition Education

Many state agriculture and education departments have extension programs providing free consulting on food purchasing and meal planning and preparation. Local chapters of the National Dairy Council also offer free advice and workshops on these subjects.

Videotape

An invaluable medium for staff training and for presenting your program to others, videotape is now being widely used by high schools and colleges, libraries, hospitals, and museums. Do not be intimidated by technology: often these groups are willing to lend videotape equipment and to train you in its use. Students can often get field-work credit for taping in your center, but, like volunteer accountants, they must be properly oriented. A half-hour of footage, which can be taped over approximately 25 times, costs about $11.

HEALTH COMPONENT IMPROVEMENT PLAN KEY*

General Instructions for Use of the Form

The Health Component Improvement Plan is a form designed to aid in problem solving and planning to improve
the health component of child care programs. The form is
divided into columns which are used for naming the problems of the health component, for specifying how changes
can be made to solve the problems, for specifying a time
frame in which the problem can be solved, and for listing
the priority rating of each problem. To use the form to its
best advantage, the following steps should be taken:

1. Naming Problems:
   Use the column headed "Health Component
   Problem" Area to name any known problems in
   each of the fifteen aspects of the health component listed on the two pages of the form. The
   problem should be named as specifically as possible to permit specific approaches to be identified.
   (See page 3 for the explanation of the content of
   each of the fifteen aspects of the health component.)
   All problems should be named before moving on
   to the next step of the problem solving process.
   Naming all the problems will permit planners to
   have a global picture of potential areas in which
   health component improvement is needed. By
   listing all the problems in each aspect of the
   health component, a more general problem may
   emerge which would not otherwise have been
   evident.
2. Naming Who is Involved in Solving the Problem:
   Using the second column on the form, identify

*Dr. Susan S. Aronson

who is involved in the solution of each of the problems named. Name the individuals affected, those with expertise, and those with the authority to bring to bear on the problem. It is possible to have a single individual who incorporates all three aspects of involvement or multiple individuals who may be affected or required for expertise for the solution of the problem. As with the naming of problems, it is helpful to identify all the individuals involved in the solution of all the named problems before proceeding to the next step in the problem solving process. In this way, it may become apparent that the same individuals are involved in the solutions to many of the problems, or conversely, that one particular problem involves a unique individual who is not involved in solving the other problems.

3.   How Changes Can Be Made:

For each problem named for which the individuals involved have been identified, attempt to define an initial process by which the problem can be addressed. The process identified may indicate a starting point or a complete approach to the solution of the problem. It is wise to attempt to specify as much detail about how the problem is to be approached as is possible from information available.

Often, the initial portion of the problem solving process is an information gathering step which will lead to further specification of an approach to the solution of the problem. As with the completion of the first two columns, the completion of the "how" column for all problems named is helpful prior to attempting to move further with the problem solving process. In this way, common approaches for solution of multiple problems will become apparent.

4. Setting a Time Plan:
   Using the information recorded in the columns headed "Health Component Problem Area", "Individuals or Agencies to be Involved in Solving the Problem" and "How Changes Can Be Made", formulate a time plan using the "Time Plan" column for each problem area named. This time plan should indicate a reasonable starting time to initiate the plan described, as well as an identification of a projected interval required to accomplish the plan described. These two aspects of the time plan should be specified clearly, but realistically. As with the other steps, complete the time plan for each of the problems named before proceeding to the next step of the problem solving process.

5. Setting Priorities:
   Using the Time Plan, your judgement of the importance of the problems named, and an awareness of the types of solutions and feasibility of solutions offered, list in priority order each of the problem areas named using the "Priority Number" column. The "Priority Number" column is used to list in order the problems to be addressed first, second, third, etc. It is not used to list the order of importance of the problems. The Priority Number is the priority assigned to working on and solving the problems. This priority can be assigned using all the information now recorded on the nature of the problems, who, how, how long, and when the problem can be solved.

General Description of Health Component Aspects Listed on the Health Component Improvement Plan Form

The following descriptions can be used to consider what, if any, problems exist in each aspect of the health component for a specific site.

1. HEALTH SERVICE SCREENING.   All children in the program should be assured of access to appropriate preventive health services. Even if the program does not directly provide any screening services, a program should assure that all children have received age-appropriate screenings either through direct arrangements or by encouraging parents to make arrangements for such care for their children.

2. HEALTH SERVICES MEDICAL EVALUATION AND TREATMENT.   As with screening, medical services need not be provided directly by the program, but assurance that adequate medical services have been provided to the children is a program responsibility. This includes the provision of adequate immunizations, medical and dental evaluation and treatment for problems identified by screening tests, and appropriate referral for mental health, nutrition, and other problems.

3. HEALTH RECORDS.   The use of summary systems aid in the maintenance of individual health records, the updating of records and the review of records. Health record data should be transferred to schools with parental consent when the child leaves the program. Health records must be maintained to assure confidentiality, but at the same time assure the use of health information by appropriate program staff. Programs should have mechanisms to assure that health information is communicated from the program to appropriate health providers at the time of both routine and special health visits. The child day care program's health record need not duplicate the detail or sophistication of a health provider record, but the child day care program's health record should document which services have been provided to each child, who provides health services routinely to the child and any services needed to achieve recommended child health standards.

4. HEALTH POLICIES AND PROCEDURES.   Written health policies and procedures should be available to help staff, parents and consultants to establish a common understanding of the principles and activities involved in the health component. There should be adequate assurance of the use of policies in staff orientation, self evaluation, and parent involvement efforts. Health policies and procedures should be reviewed annually for possible revision by staff, parents, volunteers, and consultants.

5. CHILDREN WITH SPECIAL NEEDS.   The program should assure that an adequate professional diagnosis and management plan have been obtained for each child with special needs *and* that the management plan is implemented in the day care program. Totally different programming is often not necessary, but the special needs of the child should be given consideration whenever necessary.

6. HEALTH EDUCATION.   Health education should be provided for the children, parents, volunteers, and staff including discussion, modeling, and demonstration of healthy habits and routine health procedures in the child care setting. Health education should involve the use of public service groups and community resource representatives who can supplement child care program staff's health education efforts. Health education should be conducted around the management of emergency situations, first-aid, management of minor illness, awareness of the various aspects of safety and quality of the environment, common health concerns and the rationale for any of the efforts being made to improve the program's health component.

7. STAFF HEALTH.   The program needs to assure that healthy individuals are hired and that their health status is supervised throughout their employment. Unusually frequent or long absences for illness should be investigated.

In addition, initial and on-going annual screening of staff and volunteers for infectious disease problems, assessment for chronic diseases which require special regimens or medication, and special health problems which may affect job performance should be conducted. As a part of on-going staff evaluation, staff health habits should also be assessed.

8. DENTAL CARE AND DAILY DENTAL HYGIENE.   Dental hygiene should be incorporated into daily routines. Toothbrushing should be conducted after at least one meal, with rinsing with water after sweets and snacks. Frequent use of detergent foods in menu planning also promotes dental health and establishes important habits in the child's life. Flouride supplementation should be assured where it is not already in community water supplies. Local dental health specialists and dental health education specialists should contribute to the program.

9. EVACUATION EMERGENCY AND DISASTER PLANS.   Emergency and evacuation plans and procedures should be visibly posted and frequently practiced. Each individual should be knowledgeable about his role in emergencies and should perform in such a way as to increase the efficiency of others when dealing with an emergency situation. An adequate log of drills and written emergency plans should be maintained for at least annual examination by a trained professional such as a fire marshall.

10. FIRST AID AND MANAGEMENT OF MINOR ILLNESS.   All adults relating to children in the program should bc knowledgeable about first aid methods. As a minimum, one adult per group as determined by staff/child ratios should be trained in first aid. First aid equipment and other health supplies should be available at all times and at all sites where children are in care, including on trips.

11. SAFETY AND ENVIRONMENTAL QUALITY.   Using on-site inspection, a full review of environmental quality and safety of the facility with regard to significant hazards, pest control, sanitation, ventilation, temperature, illumination, space, cleanliness, cleaning plan of the facility, environmental quality and safety surveillance routines, logging of accidental injuries, and adequacy of Health Department certification should be maintained.

12. NUTRITION.   The basic concepts of good nutrition including socio-cultural implications, nutrient quality, frequency of fluid and food intake and economic aspects of food choices should be incorporated into daily programming. The four basic food groups and variety in menu items should be modeled to children and parents. Communication through adequately prepared menus distributed in advance to parents, involvement of parents in menu planning to coordinate program meals with home meal patterns, gathering of information on feeding schedules, use of vitamin or mineral supplements, home feeding practices, food preferences, food sensitivities or allergies, are all important to providing an adequate nutrition program. Adequate supplies of food and formula should always be on hand. Group feeding routines should be comfortable socialization experiences. Appropriate and varied snacks should be served whenever possible. Staff should eat with the children.

13. MENTAL HEALTH SERVICES.   All children should receive routine assessment for developmental and behavioral problems. Mental health consultants should be involved to help staff to provide for the individual needs of each child enrolled in the program. As a minimum, mental health professionals should be involved for preventive planning and in-service education of child day care program staff.

14. TRANSPORTATION ARRANGEMENTS.   Emergency transportation arrangements need to be clearly defined and clearly communicated both to the day care program staff and to the agency supplying the emergency transportation.

Routine transportation arrangements should include adequate provision for driver education, use of safety restraints, use of safety precautions in loading and unloading arrangements and planning for pedestrian safety where vehicular transportation is not involved.

15. PLANNING AND ADMINISTRATION OF THE HEALTH COMPONENT.

    a.   External Planning. External planning includes the routine use of appropriate outside sources of expertise to assist in evaluating the entire health component and in proposing strategies for solving health related problems.

        External planning must include health professionals who are knowledgeable about community agencies and who are likely to be able to tap community resources which might otherwise be unknown to the program.

    b.   Internal Planning and Administration of the Health Component. To assure that all aspects of the health component are adequately implemented, appropriate management techniques are required. A single individual should be identified to be responsible for assuring that decisions or plans made relating to the health component are implemented. This individual may delegate the primary performance responsibility to others, but retains responsibility for assuring that all aspects of the health component are accomplished. Mechanisms need to be established to assure that health services are appropriately scheduled, that there is communication of health services and outcome schedule information between parents and program.

Each child should be assured of having some health cost coverage arrangements. These arrangements may include Medical Assistance, insurance carriers, or parental awareness and willingness to pay cash for what may become significantly large medical care costs.

Attention needs to be given to the items required to accomplish the health component which have a significant budgetary expenditure attached to them. Provision must be made within the program budget to provide for these expenses.

A routine (at least annual) re-evaluation and planning process to continue health component improvement must be established. The use of the Health Component Improvement Plan is suggested for this purpose.

This work was supported by HEW grant OCD CB 491.

INDEX

293